More Advance Praise for *Brainwashed*

"An authoritative, fascinating argument for the centrality of mind in what, doubtless prematurely, has been called the era of the brain."
—Peter D. Kramer, author of *Listening to Prozac*

"*Brainwashed* provides an engaging and wonderfully lucid tour of the many areas in which the progress and applications of neuroscience are currently being overstated and oversold. Some of the hyping of neuroscience appears fairly harmless, but more than a little of it carries potential for real damage—especially when it promotes erroneous ideas about addiction and criminal behavior. The book combines clearheaded analysis with telling examples and anecdotes, making it a pleasure to read."
—Hal Pashler, Distinguished Professor of Psychology and Cognitive Science, University of California, San Diego

"Satel and Lilienfeld have produced a remarkably clear and important discussion of what today's brain science can and cannot deliver for society. As a neuroscientist, I confess that I also enjoyed their persuasive skewering of hucksters whose misuse of technology in the courtroom and elsewhere is potentially damaging not only to justice but also to the public understanding of science."
—Dr. Steven E. Hyman, Director of the Stanley Center for Psychiatric Research at the Broad Institute of MIT and Harvard and Former Director of the National Institute of Mental Health

"There is a widespread belief that brain science is the key to understanding humanity and that imaging will X-ray our minds, revealing why we buy things and whether we are telling the truth and answering questions about addiction, criminal responsibility, and free will. *Brainwashed* is a beautifully written, lucid dissection of these exaggerated claims, informed by a profound knowledge of current

neuroscience. It is essential reading for anyone who wants a balanced assessment of what neuroscience can and cannot tell us about ourselves."
—Raymond Tallis, author of *Aping Mankind: Neuromania, Darwinitis and the Misrepresentation of Humanity*

BRAINWASHED

BRAINWASHED

The Seductive Appeal of Mindless Neuroscience

Sally Satel and
Scott O. Lilienfeld

BASIC BOOKS
A Member of the Perseus Books Group
New York

Published by Basic Books,
A Member of the Perseus Books Group

Books published by Basic Books are available at special
discounts for bulk purchases in the United States by corporations,
institutions, and other organizations. For more information, please contact
the Special Markets Department at the Perseus Books Group, 2300 Chestnut
Street, Suite 200, Philadelphia, PA 19103, or call (800) 810-4145,
ext. 5000, or e-mail special.markets@perseusbooks.com.

A CIP catalog record for this book is available from the Library of Congress.

ISBN: 978-0-465-01877-2 (hardcover)
ISBN: 978-0-465-03786-5 (e-book)
10 9 8 7 6 5 4 3 2 1

*This book is dedicated to the memory of
James Q. Wilson—scholar, gentleman, naturalist.*

CONTENTS

INTRODUCTION

Losing Our Minds in the
Age of Brain Science

YOU'VE SEEN THE HEADLINES: This is your brain on love. Or God. Or envy. Or happiness. And they're reliably accompanied by articles boasting pictures of color-drenched brains—scans capturing Buddhist monks meditating, addicts craving cocaine, and college sophomores choosing Coke over Pepsi. The media—and even some neuroscientists, it seems—love to invoke the neural foundations of human behavior to explain everything from the Bernie Madoff financial fiasco to slavish devotion to our iPhones, the sexual indiscretions of politicians, conservatives' dismissal of global warming, and even an obsession with self-tanning.[1]

Brains are big on campus, too. Take a map of any major university, and you can trace the march of neuroscience from research labs and medical centers into schools of law and business and departments of economics and philosophy. In recent years, neuroscience has merged with a host of other disciplines, spawning such new areas of study as neurolaw, neuroeconomics, neurophilosophy, neuromarketing, and neurofinance. Add to this the birth of neuroaesthetics, neurohistory, neuroliterature, neuromusicology, neuropolitics, and neurotheology. The brain has even wandered into such unlikely redoubts as English departments, where professors debate whether scanning subjects' brains as they read passages from Jane Austen novels represents (a) a fertile inquiry into the power of literature or (b) a desperate attempt to inject novelty into a field that has exhausted its romance with psychoanalysis and postmodernism.[2]

Clearly, brains are hot. Once the largely exclusive province of neuroscientists and neurologists, the brain has now entered the popular mainstream. As a newly minted cultural artifact, the brain is portrayed in paintings, sculptures, and tapestries and put on display in museums and galleries. One science pundit noted, "If Warhol were around today, he'd have a series of silkscreens dedicated to the cortex; the amygdala would hang alongside Marilyn Monroe."[3]

The prospect of solving the deepest riddle humanity has ever contemplated—itself—by studying the brain has captivated scholars and scientists for centuries. But never before has the brain so vigorously engaged the public imagination. The prime impetus behind this enthusiasm is a form of brain imaging called functional magnetic resonance imaging (fMRI), an instrument that came of age a mere two decades ago, which measures brain activity and converts it into the now-iconic vibrant images one sees in the science pages of the daily newspaper.

As a tool for exploring the biology of the mind, neuroimaging has given brain science a strong cultural presence. As one scientist remarked, brain images are now "replacing Bohr's planetary atom as the symbol of science."[4] With its implied promise of decoding the brain, it is easy to see why brain imaging would beguile almost anyone interested in pulling back the curtain on the mental lives of others: politicians hoping to manipulate voter attitudes, marketers tapping the brain to learn what consumers really want to buy, agents of the law seeking an infallible lie detector, addiction researchers trying to gauge the pull of temptations, psychologists and psychiatrists seeking the causes of mental illness, and defense attorneys fighting to prove that their clients lack malign intent or even free will.

The problem is that brain imaging cannot do any of these things—at least not yet.

AUTHOR Tom Wolfe was characteristically prescient when he wrote of fMRI in 1996, just a few years after its introduction, "Anyone who cares to get up early and catch a truly blinding twenty-first-century dawn will want to keep an eye on it."[5] Now we can't look away.

Why the fixation? First, of course, there is the very subject of the scans: the brain itself. More complex than any structure in the known cosmos, the brain is a masterwork of nature endowed with cognitive powers that far outstrip the capacity of any silicon machine built to emulate it. Containing roughly 80 billion brain cells, or neurons, each of which communicates with thousands of other neurons, the three-pound universe cradled between our ears has more connections than there are stars in the Milky Way.[6] How this enormous neural edifice gives rise to subjective feelings is one of the greatest mysteries of science and philosophy.

Now combine this mystique with the simple fact that pictures—in this case, brain scans—are powerful. Of all our senses, vision is the most developed. There are good evolutionary reasons for this arrangement: The major threats to our ancestors were apprehended visually; so were their sources of food. Plausibly, the survival advantage of vision gave rise to our reflexive bias for believing that the world is as we perceive it to be, an error that psychologists and philosophers call naive realism. This misplaced faith in the trustworthiness of our perceptions is the wellspring of two of history's most famously misguided theories: that the world is flat and that the sun revolves around the earth. For thousands of years, people trusted their raw impressions of the heavens. Yet, as Galileo understood all too well, our eyes can deceive us. He wrote in his *Dialogues* of 1632 that the Copernican model of the heliocentric universe commits a "rape upon the senses"—it violates everything our eyes tell us.[7]

Brain scan images are not what they seem either—or at least not how the media often depict them. Nor are brain-scan images what they seem. They are not photographs of the brain in action in real time. Scientists can't just look "in" the brain and see what it does. Those beautiful color-dappled images are actually representations of particular areas in the brain that are working the hardest—as measured by increased oxygen consumption—when a subject performs a task such as reading a passage or reacting to stimuli, such as pictures of faces. The powerful computer located within the scanning machine transforms changes in oxygen levels into the familiar candy-colored splotches indicating the brain regions that become especially active during the subject's performance. Despite well-informed inferences, the greatest challenge of imaging is that it is very difficult for scientists to look at a fiery spot on a brain scan and conclude with certainty what is going on in the mind of the person.[8]

Neuroimaging is a young science, barely out of its infancy, really. In such a fledgling enterprise, the half-life of facts can be especially brief. To regard research findings as settled wisdom is folly, especially when they emanate from a technology whose implications are still poorly understood. As any good scientist knows, there will always be questions to hone, theories to refine, and techniques to perfect. Nonetheless, scientific humility can readily give way to exuberance. When it does, the media often seem to have a ringside seat at the spectacle.

Several years ago, as the 2008 presidential election season was gearing up, a team of neuroscientists from UCLA sought to solve the riddle of the undecided, or swing, voter. They scanned the brains of swing voters as they reacted to photos and video footage of the candidates. The researchers translated the resultant brain activity into the voters' unspoken attitudes and, together with three political consultants from a Washington, D.C.–based firm called FKF Applied Research, presented their findings in the *New York Times* in an op-ed titled "This Is Your Brain on Politics."[9] There, readers could view scans dotted with tangerine and neon yellow hot spots indicating regions that "lit up" when the subjects were exposed to images of Hillary Clinton, Mitt Romney, John Edwards, and other candidates. Revealed in these activity patterns, the authors claimed, were "some voter impressions on which this election may well turn." Among those impressions was that two candidates had utterly failed to "engage" with swing voters. Who were these unpopular politicians? John McCain and Barack Obama, the two eventual nominees for president.

Another much-circulated study, published in 2008, "The Neural Correlates of Hate" came from neuroscientists at University College London. The researchers asked subjects to bring in photos of people they hated—generally ex-lovers, work rivals, or reviled politicians—as well as people about whom subjects felt neutrally. By comparing their responses—that is, patterns of brain activation elicited by the hated face—with their reaction to the neutral photos, the team claimed to identify the neurological correlates of intense hatred. Not surprisingly, much of the media coverage attracted by the study flew under the headline: "'Hate Circuit' Found in Brain."

One of the researchers, Semir Zeki, told the press that brain scans could one day be used in court—for example, to assess whether a murder suspect felt a strong hatred toward the victim.[10] Not so fast. True, these data do reveal that certain parts of the brain become more active when people look at images of people they hate and presumably feel contempt for them as they do so. The problem is that the illuminated areas on the scan are activated by many other emotions, not just hate. There is no newly discovered collection of brain regions that are wired together in such a way that they comprise the identifiable neural counterpart of hatred.

University press offices, too, are notorious for touting sensational details in their media-friendly releases: Here's a spot that lights up when subjects think of God ("Religion center found!"), or researchers find a region for love ("Love found in the brain"). Neuroscientists sometimes refer disparagingly to these studies as "blobology," their tongue-in-cheek label for studies that show which brain areas become activated as subjects experience X or perform task Y. To re-peat: It's all too easy for the nonexpert to lose sight of the fact that fMRI and other brain-imaging techniques do not literally read thoughts or feelings. By obtaining measures of brain oxygen levels, they show which regions of the brain are more active when a person is thinking, feeling, or, say, reading or calculating. But it is a rather daring leap to go from these patterns to drawing confident infer-ences about how people feel about political candidates or paying taxes, or what they experience in the throes of love.[11]

Pop neuroscience makes an easy target, we know. Yet we invoke it because these studies garner a disproportionate amount of media coverage and shape public perception of what brain imaging can tell us. Skilled science journalists cringe when they read accounts claim-ing that scans can capture the mind itself in action. Serious science writers take pains to describe quality neuroscience research accu-rately. Indeed, an eddy of discontent is already forming. "Neuroma-nia," "neurohubris," and "neurohype"—"neurobollocks," if you're a Brit—are just some of the labels that have been brandished, some-times by frustrated neuroscientists themselves. But in a world where university press releases elbow one another for media attention, it's often the study with a buzzy storyline ("Men See Bikini-Clad Women as Objects, Psychologists Say") that gets picked up and dumbed down.[12]

The problem with such mindless neuroscience is not neurosci-ence itself. The field is one of the great intellectual achievements of modern science. Its instruments are remarkable. The goal of brain imaging is enormously important and fascinating: to bridge the ex-planatory gap between the intangible mind and the corporeal brain. But that relationship is extremely complex and incompletely under-stood. Therefore, it is vulnerable to being oversold by the media, some overzealous scientists, and neuroentrepreneurs who tout facile conclusions that reach far beyond what the current evidence warrants—fits of "premature extrapolation," as British neuroskeptic Steven

Poole calls them.[13] When it comes to brain scans, seeing may be believing, but it isn't necessarily understanding.

Some of the misapplications of neuroscience are amusing and essentially harmless. Take, for instance, the new trend of neuromanagement books such as *Your Brain and Business: The Neuroscience of Great Leaders*, which advises nervous CEOs "to be aware that anxiety centers in the brain connect to thinking centers, including the PFC [prefrontal cortex] and ACC [anterior cingulate cortex]." The fad has, perhaps not surprisingly, infiltrated the parenting and education markets, too. Parents and teachers are easy marks for "brain gyms," "brain-compatible education," and "brain-based parenting," not to mention dozens of other unsubstantiated techniques. For the most part, these slick enterprises merely dress up or repackage good advice with neuroscientific findings that add nothing to the overall program. As one cognitive psychologist quipped, "Unable to persuade others about your viewpoint? Take a Neuro-Prefix—influence grows or your money back."[14]

But reading too much into brain scans matters when real-world concerns hang in the balance. Consider the law. When a person commits a crime, who is at fault: the perpetrator or his or her brain? Of course, this is a false choice. If biology has taught us anything, it is that "my brain" versus "me" is a false distinction. Still, if biological roots can be identified—and better yet, captured on a brain scan as juicy blotches of color—it is too easy for nonprofessionals to assume that the behavior under scrutiny must be "biological" and therefore "hardwired," involuntary or uncontrollable. Criminal lawyers, not surprisingly, are increasingly drawing on brain images supposedly showing a biological defect that "made" their clients commit murder.

Looking to the future, some neuroscientists envision a dramatic transformation of criminal law. David Eagleman, for one, welcomes a time when "we may someday find that many types of bad behavior have a basic biological explanation [and] eventually think about bad decision making in the same way we think about any physical process, such as diabetes or lung disease."[15] As this comes to pass, he predicts, "more juries will place defendants on the not-blameworthy side of the line."[16] But is this the correct conclusion to draw from neuroscientific data? After all, if every behavior is eventually traced to detectable correlates of brain activity, does this mean we can one day write off all troublesome behavior on a don't-blame-me-blame-my-brain theory of crime? Will no one ever be judged responsible? Thinking through these profoundly important questions turns on how we understand the relationship between the brain and the mind.

THE mind cannot exist without the brain. Virtually all modern scientists, ourselves included, are "mind-body monists": they believe that mind and brain are composed of the same material "stuff." All subjective experience, from a frisson of fear to the sweetness of nostalgia, corresponds to physical events in the brain. Decapitation proves this point handily: no functioning brain, no mind. But even though the mind is produced by the action of neurons and brain circuits, the mind is not identical with the matter that produces it. There is nothing mystical or spooky about this statement, nor does it imply an endorsement of mind-body "dualism," the dubious assertion that mind and brain are composed of different physical material. Instead, it means simply that one cannot use the physical rules from the cellular level to completely predict activity at the psychological level. By way of analogy, if you wanted to understand the text on this page, you could analyze the words by submitting their contents to an inorganic chemist, who could ascertain the precise molecular composition of the ink. Yet no amount of chemical analysis could help you understand what these words mean, let alone what they mean in the context of other words on the page.

Scientists have made great strides in reducing the organizational complexity of the brain from the intact organ to its constituent neurons, the proteins they contain, genes, and so on. Using this template, we can see how human thought and action unfold at a number of explanatory levels, working upward from the most basic elements. At one of the lower tiers in this hierarchy is the neurobiological level, which comprises the brain and its constituent cells.[17] Genes direct neuronal development; neurons assemble into brain circuits. Information processing, or computation, and neural network dynamics hover above. At the middle level are conscious mental states, such as thoughts, feelings, perceptions, knowledge, and intentions. Social and cultural contexts, which play a powerful role in shaping our thoughts, feelings, and behavior, occupy the highest landings of the hierarchy.

Problems arise, however, when we ascribe too much importance to the brain-based explanations and not enough to psychological or social ones. Just as one obtains differing perspectives on the layout of a sprawling city while ascending in a skyscraper's glass elevator, we can gather different insights into human behavior at different levels of analysis.[18]

The key to this approach is recognizing that some levels of explanation are more informative for certain purposes than others. This principle is profoundly important in therapeutic intervention. A scientist trying to develop a medication for Alzheimer's disease will toil on the lower levels of the explanatory ladder, perhaps developing compounds aimed at preventing the formation of the amyloid plaques and neurofibrillary tangles endemic to the disease. A marriage counselor helping a distraught couple, though, must work on the psychological level. Efforts by this counselor to understand the couple's problems by subjecting their brains to fMRIs could be worse than useless because doing so would draw attention away from their thoughts, feelings, and actions toward each other—the level at which intervention would be most helpful.

This discussion brings us back to brain scans and other representations of brain-derived data. What can we infer from this information about what people are thinking and feeling or how their social world is influencing them? In a way, imaging rekindles the age-old debate over whether brain equals mind. Can we ever fully comprehend the psychological by referring to the neural? This "hard problem," as philosophers call it, is one of the most daunting puzzles in all of scientific inquiry. What would the solution even look like? Will the parallel languages of neurobiology and mental life ever converge on a common vernacular?[19]

Many believe it will. According to neuroscientist Sam Harris, inquiry into the brain will eventually and exhaustively explain the mind and, hence, human nature. Ultimately, he says, neuroscience will—and should—dictate human values. Semir Zeki, the British neuroscientist, and legal scholar Oliver Goodenough hail a "'millennial' future, perhaps only decades away, [when] a good knowledge of the brain's system of justice and of how the brain reacts to conflicts may provide critical tools in resolving international political and economic conflicts." No less towering a figure than neuroscientist Michael Gazzaniga hopes for a "brain-based philosophy of life" based on an ethics that is "built into our brains. A lot of suffering, war, and conflict could be eliminated if we could agree to live by them more consciously."[20]

It's no wonder, then, that some see neuroscientists as the "new high priests of the secrets of the psyche and explainers of human behavior in general."[21] Will we one day replace government bureaucrats with neurocrats? Though short on details—neuroscientists don't say *how* brain science is supposed to determine human values or achieve world peace—their predictions are long on ambition. In fact, some experts talk of neuroscience as if it is the new genetics, that is, just the latest overarching narrative commandeered to explain and predict virtually all human behavior. And before genetic determinism there was the radical behaviorism of B. F. Skinner, who sought to explain human behavior in terms of rewards and punishments. Earlier in the late nineteenth and twentieth centuries, Freudianism posited that people were the products of unconscious conflicts and drives. Each of these movements suggested that the causes of our actions are not what we think they are. Is neurodeterminism poised to become the next grand narrative of human behavior?

As a psychiatrist and a psychologist, we have followed the rise of popular neuroscience with mixed feelings. We're delighted to see laypeople so interested in brain science, and we are excited by the promise of new neurophysiological discoveries. Yet we're dismayed that much of the media diet consists of "vulgarized neuroscience," as the science watchdog Neuroskeptic puts it, that offers facile and overly mechanistic explanations for complicated behaviors. We were both in training when modern neuroimaging techniques made their debut. The earliest major functional imaging technique (PET, or positron emission tomography) appeared in the mid-1980s. Less than a decade later, the near wizardry of fMRI was unveiled and soon became a prominent instrument of research in psychology and psychiatry. Indeed, expertise in imaging technology is becoming a sine qua non for graduate students in many psychology programs, increasing their odds of obtaining federal research grants and teaching posts and boosting the acceptance rates of their papers by top-flight journals. Many psychology departments now make expertise in brain imaging a requirement for their new hires.[22]

The brain is said to be the final scientific frontier, and rightly so, in our view. Yet in many quarters brain-based explanations appear to be granted a kind of inherent superiority over all other ways of accounting for human behavior. We call this assumption "neurocentrism"—the view that human experience and behavior can be best explained from the predominant or even exclusive perspective of the brain.[23] From this popular vantage point, the study of the brain is somehow more "scientific" than the study of human motives, thoughts, feelings, and actions. By making the hidden visible, brain imaging has been a spectacular boon to neurocentrism.

Consider addiction. "Understanding the biological basis of pleasure leads us to fundamentally rethink the moral and legal aspects of addiction," writes neuroscientist David Linden.[24] This is popular logic among addiction experts, but to us, it makes little sense. Granted, there may be good reasons to reform the way the criminal justice system deals with addicts, but the biology of addiction is not one of them. Why? Because the fact that addiction is associated with neurobiological changes is not, in itself, proof that the addict is unable to choose. Just look at American actor Robert Downey Jr. He was once a poster boy for drug excess. "It's like I have a loaded gun in my mouth and my finger's on the trigger, and I like the taste of gunmetal," he said. It seemed only a matter of time before he would meet a horrible end. But Downey entered rehab and decided to change his life. Why did Downey use drugs? Why did he decide to stop and to remain clean and sober? An examination of his brain, no matter how sophisticated the probe, could not tell us why and perhaps never will. The key problem with neurocentrism is that it devalues the importance of psychological explanations and environmental factors, such as familial chaos, stress, and widespread access to drugs, in sustaining addiction.

OUR goal in this book is to bring some perspective to the bold speculations surrounding the promise of neuroscience. The chapters follow the migration of brain imaging (and occasionally brain-wave technologies, such as EEG, or electroencephalography) outside the lab and medical center and into marketing suites, drug-treatment clinics, and courtrooms.

We begin in Chapter 1 with a basic overview of fMRI. We review the principles of brain organization, how scans are constructed, and how simple studies are designed. We also examine some of the potential pitfalls of interpretation introduced by brain imaging. One of our major aims is to convey an appreciation for the staggering complexity of the brain and the implications of attempts to infer mental contents, such as thoughts, desires, intentions, and feelings, from brain-derived information.

In Chapter 2 we turn to neuromarketing. The impetus behind neuromarketing is the notion that consumers are inaccurate reporters of what they truly like and plan to purchase. If their brains can be tapped to measure their immediate responses to products or other stimuli, such as commercials or movie trailers, neuromarketers, who advise many Fortune 500 companies, believe that they can guide corporations in designing the most compelling ads and sales campaigns.

The biology of pathological desire figures prominently in Chapter 3 on addiction. Indeed, within research circles and some clinical venues, the idea that addiction is a "brain disease" is the dominant conceptual framework. The mechanical simplicity of this neurocentric view carries a seductive appeal that obscures the myriad other factors that drive addiction. A broader understanding of addiction that goes beyond its biological dimension is imperative if treatment is to be successful and recovery sustained.

The remaining chapters focus on the implications of the age of neuroscience for the law. Chapter 4 examines brain-based deception detection. Like neuromarketing, it is an arena animated by a major entrepreneurial spirit. Commercial outfits, such as No Lie MRI, claim to provide security firms, employers, and suspicious spouses with "unbiased methods for the detection of deception and other information stored in the brain." Several times, No Lie and its competitor Cephos have tried to bring their evidence to court. We assess the scientific justification for using these techniques in forensic situations, where stakes are high. We also ask whether citizens are about to be confronted by the chilling words "We have a warrant to search your brain" anytime soon.

Chapter 5 on neurolaw puts neuroscience before the judge and jury. As the triers of fact consider the neurobiological facts of cases, neuroscientists such as David Eagleman and Sam Harris hope to see a general attitude "shift from blame to biology." Yet the relationship between brain-based explanations of a defendant's crime and what they mean for holding that person responsible is by no means straightforward.[25]

In Chapter 6 we explore a momentous question: What are the implications of neuroscience for individuals' freedom of choice? We generally think of ourselves as free agents who have the power to alter our destinies and earn praise or blame for our deeds, good and bad. But a number of prominent scholars claim that we are mistaken. "Our growing knowledge about the brain makes the notions of volition, culpability, and, ultimately, the very premise of the criminal justice system, deeply suspect," contends biologist Robert Sapolsky.[26] Will our coming to understand how the brain works necessitate a radically new way of thinking about human beings as moral agents worthy of blame and praise? As we will see, there is ample reason to doubt that it will.

Finally, in the Epilogue, we reprise what we have learned, and examine the crucial question of what neuroscience can—and cannot—tell us about human behavior. Brain imaging tools hold enormous potential for elucidating the neural correlates of everyday decisions, addiction, and mental illness. Yet these promising new technologies must not detract from the importance of levels of analysis other than the brain in explaining human behavior. Ours is an age in which brain research is flourishing—a time of truly great expectations. Yet it is also a time of mindless neuroscience that leads us to overestimate how much neuroscience can improve legal, clinical, and marketing practices, let alone inform social policy. Naive media, slick neuroentrepreneurs, and even an occasional overzealous neuroscientist exaggerate the capacity of scans to reveal the contents of our minds, exalt brain physiology as inherently the most valuable level of explanation for understanding behavior, and rush to apply underdeveloped, if dazzling, science for commercial and forensic use.[27]

Granted, it is only natural that advances in knowledge about the brain make us think more mechanistically about ourselves. But if we become too carried away with this view, we may impede one of the most challenging cultural projects looming in the years ahead: how to reconcile advances in brain science with personal, legal, and civic notions of freedom.

The neurobiological domain is one of brains and physical causes. The psychological domain, the domain of the mind, is one of people and their motives. Both are essential to a full understanding of why we act as we do and to the alleviation of human suffering. The brain and the mind are different frameworks for explaining experience. And the distinction between them is hardly an academic matter; it bears crucial implications for how we think about human nature, personal responsibility, and moral action.

1

THIS IS YOUR BRAIN ON AHMADINEJAD

Or What Is Brain Imaging?

I N THE SPRING OF 2008, the folks at FKF Applied Research—the political consultants and neuroscientists behind the aforementioned swing-voter study—were at it again. This time they took journalist Jeffrey Goldberg of the *Atlantic* on a guided tour of his brain. The idea had been hatched at a family Passover seder at which Goldberg had spent the evening "issuing a series of ideologically contradictory, Manischewitz-fueled political pronouncements." Fortunately for science, Bill Knapp, a political consultant and cofounder of FKF, was one of the seder guests. He suggested that if Goldberg wanted to get to the bottom of his confusion, he should submit to a brain scan to learn whether he was "neurologically wired for liberalism or conservatism." As Goldberg understood the process, researchers would measure his brain's responses to a series of images of famous politicians to uncover the truth about his "actual inclinations and predispositions by sidestepping the usual inhibition controls that can make focus-group testing unreliable."[1]

When Goldberg arrived at the facility, he was slid faceup into the mouth of a sleek MRI machine and asked to lie as still as a cadaver lest movement disrupt the readings. Despite noise-dampening headphones, Goldberg could still hear the magnet in the state-of-the-art fMRI machine as it scanned his brain, a racket that's been likened to the sound of metal-cleated golf shoes tumbling in a clothes dryer followed by a long period of high-pitched pinging.[2] The researchers had fitted him with video goggles through which they flashed scores of photographs and film clips of cultural and political celebrities, including John McCain, Edie Falco, Golda Meir, Barack Obama, Yasser Arafat, Bruce Springsteen, George W. Bush, and Iranian president Mahmoud Ahmadinejad. A lesser man might have been daunted by the fusillade of images, but Goldberg's trials as a war correspondent in the Middle East seemed to have prepared him for a full hour inside the machine. Goldberg emerged with a clanging headache, but with his sense of humor intact. "If you haven't lain supine in a claustrophobia-inducing magnetized tunnel while watching Hillary Clinton talk about health care one inch from your eyeballs, well, you just haven't lived," he quipped.

Goldberg's brain, which the fMRI declared to be nonpartisan, displayed the same ambivalent reaction to Hillary Clinton as did the swing voters.' The team's neuroscientist, Marco Iacoboni, speculated that enhanced activity in Goldberg's dorsolateral prefrontal cortex, an area linked to inhibition of ones spontaneous responses, indicated that he might be trying to "suppress unwanted emotions" about Clinton. Scanning also revealed that Goldberg loved Edie Falco by virtue of a strong response in the ventral striatum, an area of the brain that revs up at the prospect of reward. "I didn't need a million-dollar machine to tell me that," wrote Goldberg, an avowed fan of The Sopranos.

Goldberg's reaction to Ahmadinejad, however, took him by surprise. The sight of the Iranian leader also stimulated Goldberg's ventral striatum. "Reward!" Iacoboni exclaimed. "You'll have to explain this one." Although Goldberg couldn't fathom why Ahmadinejad would stimulate pleasurable thoughts, Joshua Freedman, a psychiatrist working with Iacoboni, offered a conjecture: "You seem to believe that the Jewish people endure [and] that people who try to hurt the Jewish people ultimately fail. Therefore, you derive pleasure from believing that Ahmadinejad will also eventually fail." Freedman paused. "Or it means that you're a Shiite."

Goldberg reflected on his adventure in "vanity scanning," as he called it, and questioned the analytic rigor of the procedure. "I wondered to what degree this was truly scientific and to what degree it was 21st-century phrenology." Goldberg isn't the first to express such doubts. Frustrated experts have also dubbed overeager readings of fMRI images "neophrenology" in reference to the long-discredited method of revealing a person's personality traits and talents by "reading" his or her skull's bumps and depressions.[3] Yet in many respects this analogy is unfair. Unlike phrenology, brain imaging is a technological marvel that does reveal something about the relationship between brain and mind. But exactly what can a "lit" brain region really tell us about an individual's thoughts and feelings?

This question sits at the leading edge of a large and time-honored body of inquiry: What can the workings of the brain tell us about the mind? Approaching this project through fMRI, one of the most up-to-the-minute and surely most mediagenic of neurotechnologies, hinges on scientists' ability to translate brain activity (mechanism) into accounts of what a person is thinking or feeling (meaning). Scientists, of course, cannot "read" specific thoughts with fMRI; they can only tell that brain regions already known to be associated with certain thoughts or feelings have demonstrated an increase in activity—hence the proper term "neural correlates" for the colorful dabs on brain scans. The value of brain scans in the courtroom and other venues rests on how accurately scientists can infer thoughts and feelings from these correlates. This challenging task began over a century ago, using far more primitive technologies.

NEUROIMAGING has come a long way from its earliest ancestor, the X-ray technique, invented in 1895 by German physicist William

Conrad Roentgen. His now-famous first X-ray showed the five bones of his wife's left hand, with the fourth bone encircled by a thick wedding ring. Roentgen's transformation of the previously hidden into the visible triggered a craze on both sides of the Atlantic. Department stores in Chicago, New York, and Paris installed X-ray slot machines so customers could view the skeletal anatomy of their hands, with the occasional customer fainting at the sight of his or her bones. A Parisian physician, Hippolyte-Ferdinand Baraduc, even claimed that he could use X-rays to photograph his own ideas and feelings. He called the resulting pictures "psychicons," or images of the mind. The X-ray, of course, is mute when it comes to the brain, let alone the mind, because the rays cannot pass readily through the skull's thick walls.[4]

At the turn of the twentieth century, scientists developed ventriculography, a method of pumping air into the brain's ventricles—hollow spaces that drain fluid from the brain—to increase the pressure inside and exaggerate density differentials across regions. In the early 1970s, computerized axial tomography (CT or CAT) scans enabled neuroradiologists to distinguish the white and gray matter of the brain from the ventricles that run through it. The technique uses high-density X-rays to capture images in slices and produce a three-dimensional model of the brain. A decade later, structural MRIs (magnetic resonance images) came on the radiographic scene, yielding an increasingly precise representation of brain anatomy. Structural MRI can detect static problems, such as tumors, blood clots, and deformed blood vessels. Taken together, MRIs and CT scans provide valuable information about fixed anatomy but leave us largely in the dark about the brain's functioning.[5]

That limitation began to change with the development of positron emission tomography (PET), one of the earliest three-dimensional functional imaging techniques. In contrast to structural techniques, PET and other functional methods allow neuroscientists to image the brain in action. Introduced in the 1980s, PET measures brain metabolism or brain blood flow by deploying radioactive tracer molecules. The underlying principle is that when brain cells are active, they need more energy in the form of glucose or oxygen. The tracer, typically low-dose glucose labeled with a radioisotope, is either injected directly into a vein or inhaled. The glucose travels to the most active brain cells, where it emits energy (positrons) that are detected and displayed as a glowing "hot spot" on a PET scan. Although PET can also be used to examine the brain while subjects respond to stimuli or perform tasks, neuroscientists tend to prefer fMRI for that purpose because it has higher spatial and temporal resolution and does not involve radioactive material.[6]

Functional MRI leverages the fact that everything the brain enables us to do—feel, think, perceive, and act—is linked, or correlated, with changes in oxygen consumption and regional blood flow in the brain. When a person responds to a task, such as looking at photos or solving a math problem, specific regions of the brain are typically engaged and receive more oxygen-laden, or oxygenized, blood. The increased blood flow and the boost in oxygen associated with it are proxies for increased activation of neurons. We say "increased" because the entire living brain is always on; blood is always circulating and oxygen is always being consumed. The only truly silent brain is a dead brain.

Measuring the concentration of oxygen dissolved in the blood, therefore, is the key to detecting brain activity. The large and immensely powerful magnet within the fMRI machine can measure the influx of blood to areas of the brain because blood that is carrying more oxygen has different magnetic properties than blood that has already given up its oxygen to supply neurons. The relative concentrations of oxygenated and oxygen-depleted, or deoxygenated, blood in a small area of brain tissue creates a signal known technically as the BOLD (blood-oxygen-level-dependent) response. The higher the ratio of oxygenated to deoxygenated blood in a particular area, the higher the energy consumption in that region.[7]

During experiments, researchers do not simply ask subjects to perform a task and then measure their brain activity. They measure brain activity while subjects are engaged in a task, perhaps reacting to faces, and compare it with a baseline signal present while a subject is sitting with eyes closed and mind as blank as possible. Imagine an experiment designed to identify the neural regions associated with reading aloud. Researchers ask subjects to read letters silently to themselves as they appear on a screen and then ask the subjects to read them aloud. Presumably, these two tasks engage all of the same mental processes except one. If one "subtracts" the signal generated when the subject reads silently from the signal produced when he or she reads aloud, the remaining nonoverlapping region will ostensibly be associated with speech. Brain regions engaged in common functions—such as attention and visual and mental processing of the letters, which both tasks require—will probably cancel each other out so that they appear dark in the final brain scan.

During such an experiment, the scanner's computer acquires BOLD data and collects them in tiny three-dimensional units called voxels, a portmanteau term derived from the words "volume" and "pixel." A typical brain holds about 50,000 such voxels, each measuring about three cubic millimeters. The subtraction step we described takes place on a voxel-by-voxel level. Each voxel is then assigned a color depending on the strength of the difference in activation of that individual voxel between the control and experimental conditions. The computer then generates an image highlighting the regions that become more active in one condition relative to the other. By convention, researchers use color gradations to reflect the likelihood that the subtraction—the difference in activation between the resting and the stimulated state (or between two stimulated states)— was not due to chance. The brighter the color of a region, the greater the confidence the investigator has in the difference. Thus a bright color like yellow might mean that there is only one chance in a thousand that the difference between brain activations in a given area is due to luck, whereas a darker color like purple might mean that the chances are higher, and that the brain differences are more likely to be attributable to random fluctuations in the data.

Finally, the computer filters out background noise and prepares the data to be mapped onto a three-dimensional template of a human brain. The final brain scan that we see in a magazine or on television rarely portrays the brain activity of a single person. Instead, it almost always represents the averaged results of all participants in the study. As we noted in the Introduction, any resemblance between brain scans and photographs is illusory. Photos capture images in real time and space. Functional imaging scans are constructed from information derived from the magnetic properties of blood flowing in the brain. If we removed half of the skull to observe the surface of the living brain in action, we wouldn't see a multicolored light show as various areas become active during thinking, feeling, and behaving. As striking as they are, scans are far less immediate; at their most accurate, they simply represent local activation based on statistical differences in BOLD signals.

FUNCTIONAL brain imaging is the most recent chapter in the centuries-long quest to map and comprehend the connection between the brain and the mind. Classically, the mind was regarded as the thinking part of the soul, but unlike the soul, which is by definition immaterial and is believed to survive after death, there is nothing spooky or nonmaterial about the mind. The brain enables the mind, and when the brain dies, so does the mind. The Greek physician Hippocrates, who lived around 400 BCE, is believed to have been the first to posit that the brain creates the mind. Observations of individuals with head injuries led him to conclude that "from the brain, and from the brain alone, arise our pleasures, joys, laughter and jests, as well as our sorrows, pain, grief, and tears. . . . It is the brain which makes us mad or delirious." The Epicureans of 300 BCE also believed that the human soul does not survive the death of the body. This materialist view was to be overshadowed for centuries by the dualist doctrine put forward by Hippocrates's contemporary, Plato.[8]

Plato believed that the mind, or soul, as he called it, was immortal. It floated in parallel with a person's corporeal brain, which controlled perception and movement. Components of the Platonic mind—reason, will, and desire—somehow preexisted the individual and survived after death. Plato's version of dualism prevailed more or less intact over the next five centuries until it was succeeded by the ideas of the famed Roman physician Galen, circa 200 CE. Galen posited that such faculties as memory, intellect, and imagination—that is, the rational soul—swirled within the brain's ventricles. His view was adopted by the early church fathers, who were devout dualists.

After lying dormant during a medieval interlude of several centuries, the materialist-dualist contest reawakened during the seventeenth-century French Enlightenment, when the great mathematician and philosopher René Descartes introduced another variant of dualism. Descartes was the first to advance the idea, correctly, as it turned out, that emotions, memories, and sensory perceptions are functions of the material brain. But distinct from the mechanical brain, he maintained, was a nonmaterial mind, or rational soul, that was capable of language, mathematics, consciousness, will, doubt, and understanding. The mind and the brain were connected, Descartes believed famously, through a small nub of tissue near the center of the brain called the pineal gland.[9]

Throughout the eighteenth and nineteenth centuries, anatomists and physiologists began to forge a clear association between the brain and abstract thought, emotion, and behavior. By the close of the nineteenth century, scientists, physicians, and psychologists mostly agreed that the phenomenological mind arose from the physical brain. Still, they were mystified by how the brain's chemical and electrical actions could give rise to the experience of emotional states, a problem known as the mind-brain (or mind-body) problem. Solving that problem, according to William James, the founding father of American psychology, would be "the scientific achievement before which past achievements fail." James built his own magisterial science of mental life on self-reports of his patients. Through their introspection, he developed theories of emotion, perception, imagination, and memory.[10]

As Paul Bloom notes, data suggest that even today, most adults are implicit mind-body dualists: they see the mind as largely or entirely distinct from the workings of the brain. This implicit dualism may help to explain why brain imaging studies garner so much media attention. Their results seem surprising, even fascinating, to many people. ("Wow, you mean that depression is actually in the brain? And love too?") "We intuitively think of ourselves as non-physical, and so it is a shock, and endlessly interesting, to see our brains at work in the act of thinking," remarks Bloom.[11]

Most nineteenth-century investigators relied on crude experiments to gain better insight into the human brain. Eager to apply a scientific approach, these scientists and neurologists, physicians who treated diseases of the brain and nerves, resorted to surgically destroying or deactivating parts of the brains of animals. After the operation, they observed how rabbits, pigeons, and cats moved and responded to stimuli. Similarly, to identify areas involved in sensory perception and movement control, researchers applied electric current directly to specific regions of animal brains. Research on humans, however, required either less invasive measures or nonliving subjects. By dissecting the brains of individuals who had died of head injuries, tumors, infections, or strokes, neurologists and anatomists acquired considerable insights into the relationships between anatomy, on the one hand, and emotion, intellectual function, and behavior, on the other.[12]

Possibly the most famous brain-injury case was that of Phineas Gage. A railroad foreman in Vermont, Gage lost much of his left prefrontal cortex in a grisly 1848 accident in which a long iron rod shot upward though his left cheek and flew out the top of his skull. Miraculously, Gage survived, but his former even-keeled temperament gave way to profanity, grandiosity, and belligerence. Gage's accident, later buttressed by more systematic research, helped demonstrate that the frontal lobes are the primary place, or node, at which a vast amount of neural processing comes together to regulate impulse control and social judgment.[13]

PHRENOLOGY was one of the first major brain-related theories of human behavior. During the 1800s it spread throughout Europe and the United States. Developed by Franz Joseph Gall, an esteemed Austrian-German anatomist, phrenology attempted to construct a science of brain function and human behavior. Gall believed that the mind was wholly situated within the brain. Phrenologists "read" personality by examining bumps and depressions on the skull that were supposedly responsible for dozens of traits, such as wit, inquisitiveness, and benevolence. Better-developed organs, Gall believed, pushed out areas of the overlying skull and formed bumps on the outer surface. Indentations in the cranium, in contrast, marked the weakest organs; although they had failed to grow to normal size, they could be developed through exercise, like a muscle. Accordingly, people regularly consulted phrenologists to learn their natural talents and receive advice on the type of career and life partner that would best match their brains.[14]

During a triumphant European tour between 1805 and 1807, Gall lectured to crowned heads of state, universities, and scientific societies and even received a commemorative medal from the king of Prussia—"He found a way to espy the workshop of the soul," read the inscription. Most of Gall's scientific contemporaries, however, were less enchanted. The predictive value of phrenology was dismal. Also, different examiners reading the same person's head routinely came to different conclusions about his or her personality.[15]

This is exactly what happened to Mark Twain. In the early 1870s, the great American humorist (and phrenology skeptic) had his head examined in London by the famous phrenologist Lorenzo Fowler. As

Twain described the visit in his autobiography, he was "glad of an opportunity to personally test [Fowler's] art" and disguised himself by using a fictitious name. Twain said that he was "startled" when Fowler told him that his skull bore a cavity that "represented the total absence of a sense of humor! . . . I was hurt, humiliated, resentful." Three months later, after he was certain that Fowler had forgotten him, Twain visited again—this time as himself, the famous author—and voilà, "the cavity was gone, and in its place was a Mount Everest—figuratively speaking—31,000 feet high, the loftiest bump of humor he had ever encountered in his life-long experience!"[16]

As a scheme for linking personality traits to brain anatomy, phrenology failed spectacularly, but its foundational notion—that certain types of mental phenomena are localized in the brain—is broadly correct and informs several important clinical practices today. It has become increasingly common in presurgical planning for neurosurgeons to use fMRI to map the language and motor regions of the brain to minimize damage to these functionally important areas while removing a tumor, blood clot, or epileptic tissue. Brain mapping has also been invaluable in pinpointing some of the central sites of defective activity in patients with severe, chronic depression or obsessive-compulsive disorder, allowing for optimal placement of therapeutic electrodes to stimulate the affected areas, a technique called deep brain stimulation. It is also used to ascertain stroke damage, to follow the course of Alzheimer's disease and epilepsy, and to determine brain maturity. Scientists hope that fMRI will improve the treatment of comatose patients by allowing doctors to directly measure levels of consciousness.[17]

THE idea that a specific area in the brain is solely responsible for enabling a given mental function may be intuitively appealing, but in reality it is rarely the case. Mental activities do not map neatly onto discrete brain regions. For example, Broca's area—once believed to be the brain's one and only language-production center—has been discovered not to have exclusive rights over this capacity. More precisely, it can be thought of as one of the key nodes, or convergence centers, for pathways that process language. Nor is there one designated site in charge of speech comprehension; it too relies on patterns of connectivity across multiple brain regions. Although neuroscientists regard a few cortical regions as being highly specialized for particular operations—such as the perception of faces, places, and body parts, ascribing mental states to others ("theory of mind"), and processing visually presented words—most neural real estate is zoned for mixed-use development. Furthermore, the brain can sometimes reorganize itself after injury so that other areas take over the functions of damaged regions, especially when the injury occurs early in life. For example, the "visual cortex" in blind people can be used to perceive touch, such as the feel of Braille letters.[18]

Take the variety of functions performed by the amygdala. This is a small region, one on each side of the brain, located within the temporal lobe at the point of intersection of a line that goes through the eye and another through the ear. In media reports, the amygdala has become almost synonymous with the emotional state of fearfulness. As it turns out, however, the amygdala handles much more than fear. "If I put you in a state of fear, your amygdala lights up," says imaging expert Russell Poldrack. "But that doesn't mean that every time your amygdala lights up you are experiencing fear. Almost every brain area lights up under lots of different states." Indeed, the amygdala becomes more activated during feelings of happiness, anger, and even sexual arousal—at least in women. (Perhaps, then, the women whose amygdalae lit up while looking at images of Mitt Romney circa 2007 in the swing-voter study found him appealing rather than threatening.)[19]

The amygdala also mediates responses to things that are unexpected, novel, unfamiliar, or exciting. This probably explains its increased activation when men look at pictures of a Ferrari 360 Modena. The amygdala reacts to photos of faces with menacing expressions, but also to photos of friendly, unfamiliar faces. If fearful faces are expected and happy faces are unexpected, the amygdala will respond more strongly to the happy faces. The amygdala also helps register the personal relevance of a stimulus at any given moment. One study, for example, revealed that hungry subjects manifested more robust amygdala responses to pictures of food than did their nonhungry counterparts.[20]

This example illustrates the knotty problem of reverse inference, a common practice wherein investigators reason backward from neural activation to subjective experience.[21] The difficulty with reverse inference is that specific brain structures rarely perform single tasks, so one-to-one mapping between a given region and a particular mental state is nearly impossible. In short, we can't glibly reason backward from brain activations to mental functions. When Jeffrey Goldberg views a picture of Mahmoud Ahmadinejad and his ventral striatum lights up like a menorah, some investigators might think, "Well, we know that the ventral striatum is involved with processing reward, so this subject, with his activated ventral striatum, is experiencing positive feelings for the dictator." This interpretation works only if the ventral striatum's assignment is exclusively to process the experience of pleasure. But that's not the case; novelty can also stimulate the ventral striatum.

To be fair, there is nothing wrong with the reverse-inference approach as long as the investigative buck doesn't stop there. Indeed, the approach frequently offers a valuable starting point for generating fruitful hypotheses that can later be tested in systematic experiments. Unfortunately, the studies that tend to attract media attention are the ones trafficking in conclusions based solely on reverse inference. Thus, in the swing-voter study, researchers concluded that candidate John Edwards provoked feelings of disgust among some of the undecided. Why? Because his picture spurred activity in the insula, a prune-sized area of the cortex nestled under the juncture of the temporal and frontal lobes. True, the region plays a role in mediating the experience of visceral disgust, but it too does much more. The insula has at least ten anatomical subunits, each with its own population of neurons and specialized functions that play roles in a broad range of experiences, such as trust, sudden insight, empathy, uncertainty, aversion, and disbelief. Researchers have also linked the left hemispheric insula to the quality of orgasm in women (like most brain areas, the insula is a paired structure, with one in each hemisphere). Most striking, the insula helps mediate awareness of bodily sensations. It integrates visceral states, such as pain, hunger, thirst, and temperature, and thereby contributes to the conscious experience of emotion.[22]

So, did the amygdala-energized swing voters viewing Mitt Romney experience anxiety or a sense of novelty? Or something else? Were the insulae of undecided voters signaling an attraction to John Edwards or a repulsion? Is Jeffrey Goldberg pro-Israel or a closet Shiite? The most daring neuropundits, as journalist Daniel Engber has labeled them, are not humbled by these complexities. They seem to regard a brain image almost as a kind of high-tech Rorschach inkblot test. But reading into largely ambiguous patterns what one wants to see is a serious breach of the fundamental test of a good theory: falsifiability, the ability of a hypothesis to be disproved by a test or observation.[23]

WHENEVER a newspaper headline proclaims, "Brain Scans Show . . . ," the reader should entertain some healthy skepticism. There are several reasons why.

First, brain scans rarely allow investigators to conclude that structure X "causes" function Y. This is not what fMRI alone can demonstrate. Instead, it at best indicates only correlation—that is, which parts of the brain are active when a person participates in a particular task—not which brain area is causing a particular psychological operation or behavior. For example, some teenagers' brains show increased activity in regions associated with aggressive tendencies when they play violent video games. But we cannot conclude from this observation alone that violent videos cause violent behavior. That inference would be unfounded. Perhaps teens with known aggressive tendencies also enjoy playing those games.[24] Or maybe teens whose parents are generally inattentive to their doings—including playing violent games—set the stage for all manner of misbehaving by their kids. And what about teens who are otherwise well behaved but just happen to be stimulated by these videos when they are playing them?

Second, the subtraction technique used in most fMRI experiments is not necessarily well suited to the question being asked. Recall that the assumption behind subtraction is that two mental tasks' conditions differ in only a single cognitive process. However, most mental operations that seem like a unified task are composed of many smaller components. Consider what is involved in performing a simple arithmetic problem. First, subjects must recognize a visually presented number. Second, they must register its numerical magnitude. Third, they must compute the correct solution to the problem at hand. Because these operations are not performed by the same brain region, researchers must "decompose" the math task into the neural correlates associated with each step.[25]

Now, if decomposing an arithmetic task is complicated, imagine the difficulty of breaking down even more complex states, such as attitudes and emotions. Is it even possible to translate a complex pattern of neural activity into a simple interpretation, as the swing-voter team did? Was it reasonable to infer that swing voters were "trying to suppress unwanted emotions" while they looked at Hillary Clinton? There is good reason to be dubious.

Third, although neuroimaging has deepened our knowledge of brain anatomy and function, its popular application tends to reinforce the misbegotten notion of the brain as a repository of discrete modules that control distinct capacities to think and feel. This is not the case, of course. Studies that suggest a "brain spot for X" are typically misleading because mental functions are rarely localized to one place in the brain. There is a babel of crosstalk among numerous brain regions as they are strung together in specialized neural circuits that work in parallel to process thoughts and feelings. Almost nothing is static in the brain. The organ continuously rewires itself in response to experience and learning by altering the strength of its connections countless times every second. Neuroscientists now think of the brain as an ever-changing ecosystem crackling with electrochemical energy from which our thoughts, emotions, and intentions arise, rather than a collection of blinking neural islands.

The heavily interconnected aspect of the brain explains why researchers are increasingly moving away from the kind of regional brain mapping used in the swing-voter study and embracing an fMRI technique called pattern analysis. Pattern analysis, also called decoding, mathematically examines the brain's extensive interconnections. Investigators first gather data on the "correct" brain response—say, fear—as it is observed in subjects asked to view frightening things. Once the computer program has been "taught" to determine what the subject is looking at, researchers can later infer what he or she is watching just by analyzing activity across the brain. Thus, instead of inferring that a photo of Mitt Romney induces anxiety, researchers might collect patterns of brain activity evoked by known anxiety inducers (photos of spiders, snakes, and hypodermic needles, perhaps) and then see whether the pattern that Romney elicits is a statistical match with the pattern for anxiety derived from the inducers.[26]

A fourth caveat to keep in mind when one is interpreting brain scans is the importance of experimental design. The way in which investigators design their task can exert a big impact on the responses they obtain. This lesson is powerfully illustrated by a set of experiments designed to explore how teenagers process emotional information compared with adults. In a 1999 fMRI study, Harvard University researchers asked normal teens to look at a series of black-and-white photos depicting frightened faces. They mischaracterized one in four photos, sometimes seeing anger, surprise, confusion, or even happiness. Regardless of whether they correctly identified the emotion as fear, subjects displayed significant amygdala activation. A subsequent study showed that adults made few errors when identifying fear. "I think this has important implications in terms of . . . trying to inhibit their own gut responses," said one of the researchers.[27] The studies were taken to mean that teens were innately deficient in their ability to interpret the emotions of others in social situations and, therefore, more prone to impulsive violence than adults. In Chapter 5, we show how defense lawyers have employed these kinds of findings to argue that teens are less criminally responsible for murder than are adults.

It turned out, however, that teens might not be so bad at detecting fear in others after all. When one of the original Harvard investigators, Abigail Baird, conducted additional trials with new photographs of faces, she obtained different results. She switched the photos from the dated black-and-white shots of people who resembled bad actors in a B-grade horror film (the ones used in the first two studies) to color photos of more contemporary-looking people. When she did this, Baird's adolescents provided correct responses almost 100 percent of the time. "They were simply more engaged by more contemporary peer photos in color," Baird concluded. "They did well if they cared."[28] Whatever elements of the new stimuli lead to the differences in teens' responses, the point is that a seemingly trivial aspect of the experimental task—aspects of the photo unrelated to fear—gave rise to an entirely different conclusion about the ability of teens to identify the facial expression of fear.

A fifth caveat stems from the fact that fMRI is an indirect method. Contrary to popular belief, imaging does not measure action of brain cells per se. True, most neuroscientists regard BOLD as a reasonable proxy for changes in neuronal activation, but the link between blood flow and neural activity is not straightforward. For example, there is a delay of at least two to five seconds between activation of neurons and the increase in oxygen-rich blood flowing to them. Thus information about mental processes occurring in the brain may be out of sync with the neural activity actually producing it, and therefore, rapid fluctuations of neural activity may go undetected. To compensate for lost data, researchers use electroencephalography (EEG), which detects electrical activity on the surface of the brain very rapidly, producing a data point every four milliseconds (thousandths of a second) or less, thousands of times faster than fMRI creates a single brain image.[29]

But even when neural activity is detected, it is not always clear what is going on in the hot spot. When brain cells, or neurons, fire, they send electrical impulses down a long, tapered filament called an axon. When the impulse reaches the end of the axon, neurotransmitters, or chemical messengers, are released into a synapse—a tiny gap between the axon and another neuron—and influence the action of the receiving neuron. The chemical message released by the firing neuron often stimulates the receiving neuron, making it fire. But although some neurons are excitatory, ramping up the activity of certain brain regions, others are inhibitory, tamping them down. As a consequence, regions of inhibitory neurons may "light up" brightly in the final scan, even though they are working to depress rather than stimulate activity elsewhere in the brain.

Alternatively, spots may be dark when they should show activity. This might happen when the voxel, at three cubic millimeters, is still too big a unit of spatial resolution to capture activity occurring on a smaller scale, such as in a tiny cluster of neurons that, despite its size, happens to perform a critical function. These small clusters may or may not appear on the final scan. What's more, a given region may appear deceptively less engaged with a task, on the basis of activation levels, than it really is. In fact, the region could be very important to enabling the task but appear less active because the brain becomes more efficient at tasks it performs repeatedly or automatically. Such a "practice-suppression" effect means that the blood oxygen level required to perform the task is lower than it would be for someone who has never before performed it. It's essential, therefore, to take practice effects into account when one is gauging the relative contribution of various regions.[30]

Last, it is important to keep in mind that before the final data even "reach" the voxel, analysts must deploy statistical approaches to extract meaningful information from the noise. This is where, as imaging expert Hal Pashler puts it, this "hellishly complicated [process] creates great opportunity for inadvertent mischief." The mischief is not intentional, of course. It is due partly to the fact that analytic methods are constantly evolving and can vary from lab to lab. Such lack of standardization, not unexpected in a rapidly growing field, has implications for reproducing the work of others, collaborations across labs, and building on the work of other teams.[31]

Additional mischief involves statistical error, not the neuroimaging process itself. When researchers run large numbers of statistical tests simultaneously on the BOLD signals, some of those tests are bound to turn up "statistically significant" just as a result of chance. In other words, these results will suggest, by mistake, that the brain is more active when a subject performs a task when, in fact, that part of the brain was actually not recruited. To make this point in a dramatic way, neuroscientist Craig Bennett set out to demonstrate how brain scans can produce (literally) fishy results. Bennett and his team purchased a dead Atlantic salmon from a store, placed their cooperative subject in a brain scanner, "showed" it photographs of people in various social situations, and "asked" the salmon to guess what the people were feeling. Bennett's team found what they were looking for: a tiny area in the salmon's brain flared to life in response to the task. Of course, this island of brain activation was merely a statistical artifact. Bennett and his fellow researchers had deliberately computed so many subtractions that chance alone caused a few of the results to become statistically significant despite their being entirely spurious.[32] The salmon "study," which won a 2012 Ig Nobel

Prize (for work "that makes people laugh, then think"), illustrates that decisions in data analysis can impact the reliability of fMRI results.

It is relatively easy to correct for the problem of false positives by using standard statistical tests. But there are plenty of other pitfalls. In a "bombshell" paper, as a fellow neuroscientist put it, MIT graduate student Edward Vul concluded that something was deeply wrong with how many brain-imaging researchers were analyzing their data.[33] Vul had become suspicious when he came across what he described as "impossibly high" estimated relationships between psychological states and the activation of various brain regions. Vul was skeptical, for example, of a 2005 study that purported to find a near-perfect relationship of .96 (with 1.0 being the maximum) between proneness to anxiety in response to angry speech and activity in the right cuneus, an area at the back of the brain believed to be involved in impulse control. Another implausible finding came from a 2006 study that reported a correlation of .88 between self-reported partner jealousy over emotional infidelity and activation in the insula.

In poring over the original articles, Vul and his collaborator Hal Pashler realized that these researchers were drawing inferences from a biased sample of results. When investigators look for correlations between stimuli and brain activations, they often cast a wide net. This leads them, first, to tiny regions of highest activation. Once they have homed in on those small regions, researchers compute the correlation between the psychological state in question and brain activation. In doing so, they are inadvertently capitalizing on chance fluctuations in the data set that are unlikely to hold up in later studies.[34]

Many aspects of Vul's critique are technical, but his basic point is easy to grasp: If you search a huge set of data—in this case, tens of thousands of voxels—for associations that are statistically significant and then do more analyses on only those associations, you are almost guaranteed to find something "good." (To avoid this mistake, the second analysis must be truly independent of the first one.) This error is known variously as the "circular analysis problem," the "noninde-pendence problem," or, more colloquially, "double-dipping."

These are not arcane matters. The strength of the correlation tells future researchers how to design their investigations; it alerts them not only to where they ought to look, but also to where they shouldn't. When the article of Vul and his colleagues detonated in the research community, some of the criticized authors pushed back. Counter-charges, rebuttals, and rebutted rebuttals flew around the web.[35] Yet, in the end, most scientists agreed that the statistical problem Vul identified had troubling implications, and that caution was indeed warranted in moving forward.

FUNCTIONAL brain scanning plays a vital role in the nascent study of human brain-behavior relationships. To fully appreciate its virtues and limits, one must keep three general points in mind. First, even the most superficial insight into how brain scans are generated should dispel the assumption of naive realism—the commonsense theory of perception we discussed in the Introduction. Naive realism, as phi-losophers define it, is an intuition held by most people that the view of the world that we derive from our senses is to be taken at face value. Scans show how profoundly we will be misled if we view them through the lens of naive realism. They are not raw snapshots of the brain's real-time functioning. They are highly processed representa-tions of the brain's activity.[36]

University of Montreal researcher Eric Racine refers to a corollary assumption termed "neurorealism." A first cousin of naive realism, neurorealism denotes the misbegotten propensity to regard brain im-ages as inherently more "real" or valid than other types of behavioral data. Neuroeconomist Paul Zak has described his work on the neu-robiology of trust: A brain scan "lets me embrace words like 'moral-ity' or 'love' or 'compassion' in a non-squishy way. These are real things." So is the fact that we like to talk about ourselves—but it is not really news. In discussing the psychological impact of combat, a researcher quoted as saying that brain imaging tells us that post-traumatic stress disorder (PTSD) is a "real disorder" (as if we did not know this). We coined the term "neuroredundancy" as a "neurologism" to denote things we already knew without brain scanning.[37]

A second crucial consideration is the design of the experiments. The kind of task that researchers put before their subjects—whether it involves showing them photos of loved ones or fails to account for color in photos of scared people—can have a profound influence on the neural correlates that appear in the final scan. Even seemingly trivial differences in the nature of the experimental setup can generate huge differences in brain-imaging findings. When it comes to interpreting the results of imaging studies, context is everything.

Third, a healthy sense of caution should be aroused when one hears of a study in which pat psychological explanations were derived from brain activity. There is a big difference between an investigator reporting that "region A exhibited enhanced activation when subjects viewed candidate B but not candidates C and D"—the accurate interpretation—and the less cautious conclusion that "the activity in region A means that voters prefer candidate B over others," or, worse still, "the activity in region A means that voters prefer candidate B over others because they find him to be sexier [or friendlier, more appealing, smarter, or whatever]."

BEARING these caveats in mind may help to restrain premature enthusiasm regarding the promise of brain imaging. When fMRI, PET, and other brain-imaging technologies first gained widespread currency in the heady days of the 1980s and 1990s, scores of scientists—perhaps not heeding these cautions sufficiently—confidently predicted a revolution in our understanding of mental illnesses, addiction, emotions, and personality. The scientific potential of what former president George H. W. Bush anointed "The Decade of the Brain" on July 17, 1990, seemed almost limitless. The fields of neuroscience, psychology, and psychiatry, many sensed, were on the brink of a new paradigm.[38]

Forecasts from the field's most esteemed leaders were expansive. In her 1984 book *The Broken Brain—The Biological Revolution in Psychiatry*, psychiatrist Nancy C. Andreasen, who went on to receive the National Medal of Science from President Clinton, predicted that "as they improve and become more accurate, these imaging techniques and other laboratory tests for mental illness will become part of standard medical practice during the coming years, thereby improving the precision of diagnosis and assisting in the search for causes."[39] Two years later, Herbert Pardes, then head of the National Institute of Mental Health, wrote that "neuroscience is offering not only new information but startling new technologies and approaches. . . . While much in the way of clinical implications from brain research is promise, there is an expectation of great change over the next ten to twenty years."[40]

Comparing the views of the current head of the National Institute of Mental Health, psychiatrist Thomas Insel, with those of Herbert Pardes over two decades ago, is instructive. As Insel observed in a sobering 2009 article, there is no evidence that the past two decades of advances in neuroscience have born witness to decreases in mental disorders' prevalence or to any impact on patient life span.[41] The failure of brain-imaging techniques to have yet made major inroads into the causes and treatment of mental illness offers a necessary reminder for modesty in our expectations.

Yet such modesty is not evident in all quarters. In a worrisome recent development, a popular nationwide outfit called Amen Clinics promises patients that it can diagnose and treat depression, anxiety, and attention-deficit hyperactivity disorder using brain scans. Its founder, psychiatrist Daniel Amen, oversees an empire that includes book publishing, television shows, and a line of nutritional supplements. Single photon emission computed tomography, SPECT, a nuclear-imaging technique that measures blood flow, is the type of scan favored by Amen. His clinics charge over three thousand dollars for an assessment and, according to the *Washington Post*, grossed about

$20 million in 2011. Despite near-universal agreement among psychiatrists and psychologists that scans cannot presently be used to diagnose mental illness, Amen insists that "it will soon be malpractice to not use imaging in complicated cases," as he told a symposium at an American Psychiatric Association meeting.[42]

Brain-imaging experts, however, shy away from such expansive claims. They are well versed in the conceptual limits of inferring mental states from biological indicators. They readily understand that the prowess of fMRI, a remarkable technology that is still relatively young and destined to evolve, is best demonstrated in the cognitive or affective neuroscience lab.[43] The danger comes when scanning leaves the experimental realm for socially consequential domains such as law and business, where much-needed interpretive restraint often gives way to extravagant claims about what brain scans can tell us about the mind.

Nowhere is this truer than in the nascent field of neuromarketing, our next topic, where brain science meets advertising hype. Savvy neuroentrepreneurs are selling brain-scanning and other techniques to corporate clients with the promise of unlocking the mysteries of consumer buying behavior. Using brain science to enlarge on valid insights into consumer behavior is one thing—indeed, it is a serious pursuit by a growing cadre of academics—but the most brazen promoters of neuromarketing may be pulling off little more than a brain scam.

2

THE BUYOLOGIST IS IN

The Rise of Neuromarketing

"**T**HE ULTIMATE NO-BULLSHIT ZONE.**" This is how globe-trotting Danish branding expert Martin Lindstrom refers to the human brain. "Our truest selves react to stimuli at a level far deeper than conscious thought," he writes, estimating that a whopping 90 percent of our buying decisions take place at this level. As a result, "we can't actually explain our preferences, or likely buying decisions, with any accuracy." The author of the 2008 business bestseller *Buyology: Truth and Lies About Why We Buy* and one of *Time* magazine's top 100 "Scientists and Thinkers," Lindstrom advises marketers to cut out the middlemen—the buyers themselves—and ask their brains directly: Will you buy our product? Forget focus groups and questionnaires. The brain is the route to the heart's desire.[1]

Lindstrom is a high-profile member of an upstart generation of Mad Men known as neuromarketers. They apply the tools of neuroscience, such as fMRI and brain-wave technologies, to learn how consumers' brains instantly react to ads and products. It's all in the service of answering elusive questions as old as advertising itself: What do customers want? What motivates them to buy? And how can I get them to buy *my* product? "Half the money I spend on advertising is wasted. The trouble is I don't know which half," Gilded Age department-store magnate John Wanamaker famously said. His lament still echoes today. American businesses spend billions on advertising each year—$114 billion in 2011. Yet, according to marketing experts, 80 percent of all new products either fail within six months of launch or fall significantly short of their profit forecasts.[2]

Corporations such as Google, Facebook, Motorola, Unilever, and Disney have hired neuromarketers to help them improve those odds. Has this hiring paid off? It's hard to know. Neuromarketing is a controversial practice without an established track record. Many of its purveyors lean heavily on hype. One buyologist—we use the term "buyologist" to denote marketers who routinely exaggerate what neuroscience can do to sell widgets—is A. K. Pradeep, head of the U.S. firm NeuroFocus, which, Pradeep says, can offer its corporate clients "secrets for selling to the subconscious mind." FKF Applied Research (which sponsored the notorious swing-voter study described in the Introduction) touts its "scientifically sound, empirically precise brain scan approach." To the nonexpert, neuromarketing seems able to drill down to the physiological essence of desire. Consumer choice "is an inescapably biological process," claims Neuroco, a neuromarketing firm located in the United Kingdom.[3]

The media routinely abet the mystique. "They mine your brain so they can blow your mind with products you deeply desire," gushed a 2011 article in the business magazine *Fast Company*. When reporters first began to cover neuromarketing around 2004, the consumer's "buy button in the brain" was a favorite metaphor. Other versions of a discrete buying "center" now animate the small army of neuromarketing boosters—coaches, consultants, and workshop leaders—that has formed. A company called SalesBrain, for example, touts its ability to show marketers how to "maximize your ability to influence the part of the brain that decides: the Reptilian Brain. . . . You will walk away [from the seminar] with a clear and simple methodology that brings proven science to the act of selling and persuasion."[4]

Claims such as these led the prestigious journal *Nature Neuroscience* to editorialize in a 2004 column titled "Brain Scam?" that "neuromarketing [might be] little more than a new fad, exploited by scientists and marketing consultants to blind corporate clients with science." Even friendly critics say that it is hard to judge the rigor and value of neuromarketing in the absence of clear and detailed documentation of the complex methods and research protocols that neuromarketers use. Still, the fact that an impressive cohort of esteemed scientists have joined the advisory boards of various neuromarketing companies—one even boasts a Nobel Prize winner in medicine—suggests at least some kernel of promise in the neuromarketing enterprise.[5]

By conventional metrics, neuromarketing has not yet penetrated all that deeply into the advertising world. A 2011 survey of almost seven hundred marketing professionals revealed that only 6 percent used imaging and brain-wave analysis in client work. Yet *Advertising Age*, the leading industry publication, has speculated that the use of neuromarketing by some of the biggest names in consumer goods "suggests that the early adopters are seeing results." Perhaps companies are indeed seeing results, but the evidence is largely under wraps. Firms do not publish their research, both to keep contractual agreements with their clients and to protect their own proprietary methodologies and mathematical algorithms. As a result, few detailed case studies of neuroscience influencing real marketing decisions by named companies are publicly available for review. "Until there are publications in peer-reviewed journals, there will always be a whiff of pseudo-science surrounding neuromarketing," admits Roger Dooley, host of a well-regarded blog on the subject.[6]

That whiff of pseudoscience is a common problem that plagues many efforts to apply brain imaging and other brain-based technologies outside the lab and the clinic. In this respect, neuromarketing is a microcosm of the far broader tendency within popular neuroscience to engage in neurohype. At its worst, neuromarketing succumbs to the kinds of errors in interpretation, such as reverse inference, neurocentrism, and neuroredundancy—using brain science to demonstrate what we could find out more simply by asking people directly—that can give brain imaging an undeservedly bad name. And when profits are involved, the threshold for playing fast and loose with brain science may be lowered even more.

Lindstrom, for example, has made headlines by proclaiming that the brains of Apple product users show neural patterns identical to those displayed by the brains of devoted Christians viewing a religious figure or icon. (Is it a coincidence that Lindstrom routinely advises his corporate clients to "treat your brand as a religion"?) Later, Lindstrom claimed that iPhone users are "in love" with their phones because, like love, the gadget activates the insula; never mind that the structure mediates other emotions as well. Furthermore, neuromarketing lapses too readily into neurocentrist cheerleading. Although there is little debate among cognitive psychologists that immediate emotional responses operating outside awareness influence many of our decisions, neuromarketers often take this conclusion too far. They relentlessly drive home the debatable point that immediate neural responses are inherently more authentic and predictive of consumer behavior than is conscious reflection.[7]

SINCE the turn of the twentieth century, businessmen have sought the advice of psychological experts to unlock the secrets of the consumer mind. In the 1920s, the influential American psychologist John B. Watson promoted a basic learning theory of advertising: Consumers buy a product when they have an incentive to do so. One surefire way to cultivate this desire, Watson advised companies, was to appeal to people's self-image—and the emotions and cultural associations that went with it.[8]

As a behaviorist, Watson famously treated the mind as a "black box." He was not interested in its inner workings, only its behavioral outputs. But the idea that the consumer has a strong irrational streak that marketers need to harness was persistent. Melvin Copeland's 1924 textbook *Principles of Merchandising* attributed buying behavior to both rational and irrational drives. "Motives have their origins in human instincts, and emotions represent the impulsive or unreasoning promptings to action," he wrote.

The idea that most of our actions, desires, and fantasies contain hidden meaning created a niche for Freudian theory in marketing. By the 1930s, a psychodynamic model of the consumer mind came to the fore, embodied in the writings of Ernest Dichter, an ambitious émigré from Vienna who arrived in America in 1938.[9] "You would be amazed to find how often we mislead ourselves, regardless of how smart we think we are, when we attempt to explain why we are behaving the way we do," Dichter observed. He developed a system called "motivational research." Trained interviewers administered Rorschach inkblot tests and "depth" interviews in which participants free-associated to products, and investigators then examined their narratives for themes of Freudian conflict, sex, and aggression. Dichter is perhaps best known for advising General Mills to design a cake mix that required an egg for its Betty Crocker cake mix, partly to assuage the housewife's unconscious guilt for taking a baking shortcut by using a mix, and partly because the egg symbolized a fertility offering to her husband.[10]

With the darkness of the Depression era lifting after World War II, household austerity gave way to a marketer's bounty. Although Madison Avenue was no less interested in manipulating the consumer, it was becoming increasingly disillusioned with the ability of Freudian theory to predict buyer behavior. By the mid-1960s, most agencies had abandoned the analytic approach because they found it too unscientific and its sensational claims unfulfilled.[11]

Madison Avenue was already turning to a more straightforward approach: market research. Rather than attempting to uncover customers' secret motivations, it simply asked them what they thought of a product and whether they would buy it. Focused group interviews (not known as "focus groups" until the late 1970s) combined the interview approach with polling. The groups typically consisted of a dozen or so individuals, mainly homemakers, who were led by a professional moderator in a free-form but comprehensive discussion of their reasons for liking a product, ad, radio spot, or commercial. On the basis of participants' enthusiasm, or lack thereof, executives decided to kill a product, modify it, or push it further down the production pipeline.[12]

Although focus groups remain a useful method in electoral politics and public opinion research, they have notorious weaknesses. Sometimes, a single forceful participant may sway or intimidate the others in the group. Participants often tell moderators what they think the moderators want to hear, rather than giving truthful answers, or censor their true reactions so they will fit in with the group.

A deeper issue, though, is the premise that group participants are valid informants. "The groups were basically a waste of time because there is often such a tenuous relationship between a participant's expressed intention to purchase and actual buying behavior," explains Gerald Zaltman, an emeritus professor at the Harvard School of Business who once ran such groups. The typical participant knows what she likes but not *why* she likes it or, more crucially, whether she will buy the featured product. This is because, as Dichter and earlier consumer psychologists recognized, decisions are shaped by a multitude of factors, many of them operating outside awareness, such as past experiences and personal and cultural influences that would take too long to consider individually.[13]

This insight led advertisers to the lab, where they tried to measure consumers' physiological responses to advertising. In the early 1960s, researchers experimented with pupillometry, or measures of spontaneous pupil dilation, to gauge interest in features of package designs or print advertisements. (Of course, pupillary dilation can reflect anxiety, fear, or stress, as well as interest.) They examined skin conductance response, a measure of the sweatiness of the palms, as an indicator of people's emotional response to advertisements and employed eye tracking to reveal where on a page or TV screen people's eyes traveled. In the 1970s, researchers first used electroencephalography (EEG), which measures the electrical activity of the brain by means of electrodes placed on the scalp, to examine left- and right-brain activations in response to marketing stimuli. A decade later, they added steady-state topography (a cousin of EEG that is highly sensitive to the speed of neural processing) to ascertain whether long-term memory encoding during advertising is linked to changes in consumers' preferences for certain brands. In the end, however, experts did not find these approaches particularly revelatory.[14]

Within the past two decades, refinements in brain-wave technology (primarily EEG) and the advent of brain-imaging technology have revived a biological approach to the consumer mind. Zaltman, sometimes called "the Father of Neuromarketing," conducted some of the earliest studies using PET scans in the 1980s and fMRI a decade later. From his Mind of the Market lab at Harvard Business School, Zaltman and colleagues showed advertisements and products to subjects to evoke neural patterns related to emotion, preference, or memory. In one study, the team scanned subjects while half of them examined a detailed cartoon sketching for an ad and the other half looked at the finished ad as it might appear in a magazine. Neural activity was comparable under both conditions, which prompting the team to suggest that the client need not take the very expensive step of going from the artists' rendering to a finished ad. In 1999, British neuroscientist Gemma Calvert established Neurosense in Oxford, England, the first company to apply brain imaging to consumer psychology. In the United States, the Atlanta-based BrightHouse Neurostrategies Group was established in 2002.[15] Such corporate giants as Coca-Cola, Home Depot, and Delta Airlines were among BrightHouse's earliest clients.

NEUROMARKETING, it turns out, is a big tent. There are the buyologists who hype their wares. Then there are their better-behaved counterparts: neuromarketing firms that are more circumspect—and less visible to the lay public. Annie Lang, an Indiana University psychophysiologist, was part of a 2011 panel convened by the nonprofit Advertising Research Foundation that reviewed the methods used by a number of neuromarketing firms, including the more understated ones—notably, NeuroFocus declined to participate. In Lang's words, "The claims of a couple of companies were reasonable. Their measures seemed valid; they did good statistical testing, delivered good inferences, and they were appropriately cautious with their conclusions." Last, there are the academics. They are interested in the cognitive and neuroscientific underpinnings of preference formation and of decision making. They do not call themselves neuromarketers, but their work is routinely invoked as the conceptual foundation of neuromarketing. Prominent among these scholars is the Nobel Prize winner Daniel Kahneman. By refining theories of how emotion and cognition operate in an economic setting, Kahneman and his late collaborator Amos Tversky have greatly enriched our understanding of consumer psychology.[16]

In a now classic series of experiments performed in the 1970s, Kahneman and Tversky explored how people make decisions. Fusing psychology and economics in what is now called behavioral economics, they identified certain "cognitive biases," largely unconscious errors of reasoning that distort our judgment of the world. They also pinpointed several "heuristics," mental shortcuts that help us conserve cognitive energy but that can also yield surprisingly irrational and suboptimal results in certain situations. Typical of these shortcuts is loss aversion: our tendency to care much more about avoiding loss than about accumulating gain. This discovery carries a potent subtext: Deriving satisfaction from transactions is not solely a matter of how much financial value an individual gains; it also involves how effectively that individual can reduce the anxiety that accompanies the prospect of loss. Another important cognitive bias is framing, a phenomenon whereby people tend to respond differently to the same information depending on how it is presented. For example, patients are more likely to accept a treatment that will give them a 90 percent chance of surviving over one that carries a 10 percent chance of dying. The twist is in how the options are presented: A high probability of living sounds better than a low probability of dying, even though the options carry equivalent probabilities.[17]

Refracting choice through the lens of cognitive psychology captures human behavior better than the once-prevailing assumption that consumers are rational creatures who always balance costs and benefits to serve their economic interests. Based on his findings with Tversky, Kahneman elaborated upon the concept of two independent systems that influence judgment in the face of uncertainty. The first, System 1, is responsible for split-second, intuitive, emotional thought processes. It operates with little effort or sense of voluntary control. System 2, by contrast, is the source of slow, logical, and skeptical thought. It can diminish the emotional intensity of a response and pave the way for a more deliberative appraisal. System 2 asserts itself as one mulls the pros and cons of, say, buying regular versus extrastuffed Oreos or a convertible instead of a hardtop.

With its instant access to a vast store of emotional memories and time-saving habits of cognition, System 1 enables people to make rapid judgments. When consumers' choices differ from their stated preferences, the discrepancy may well be due to dynamics operating below the threshold of awareness. For students of consumer behavior, understanding—and potentially controlling—System 1 holds greater promise. Indeed, neuromarketing firms have capitalized on the narrative of buried truth: Lucid Systems positions itself as "your source for the unspoken truth." The head of Neurosense says that he wants to "look inside the black box [of] the brain to get at insights that focus groups can't begin to explain."[18]

ATTEMPTS to illuminate the "black box" have spurred academic collaboration among economists, neuroscientists, and consumer psychologists. Neuroscience has much to say about such major phenomena as attention, emotion, and memory that are essential to motivating consumers. In 2008, neuroscientist Hilke Plassmann and colleagues published a now-famous wine-tasting experiment designed to elucidate the neural mechanisms linked to the phenomenon of framing. When people think that they are drinking a $50 bottle of wine instead of a $5 bottle—even when it's the same stuff—their brains manifest the kinds of neural patterns that are well established as associated with the experience of pleasure. Thus, when researchers scanned subjects who sipped pricey wines, the participants' brains showed increased activity in the medial orbitofrontal cortex, an area involved in the regulation of emotions and in encoding the "value" of experiences. In contrast, brain regions related to perception of taste remained insensitive to price changes. The researchers reasonably hypothesized that an item's price does not directly change the sensory experience, but rather leads people to think that the experience of consuming it is more valuable. In an experiment like this, the behavioral outcome is not news—it has been demonstrated many times already—rather, the point is to dissect the decision making apparatus at the level of the brain. (In price-blind taste tests, people don't like more expensive wines any better than less expensive ones.)[19]

In 2004, neuroscientist Read Montague reported another much-cited exploration of consumer preference that focused on the neurobiology of branding by using the famous Coke-Pepsi challenge. Montague and his team asked why Coke consistently dominates the market even though in blind taste tests subjects tend to prefer Pepsi or have no reliable preference for one cola over the other. The investigators put subjects in an fMRI scanner, where they received random "blind" sips of Coke and Pepsi through long straws, not knowing which brand they were given. When subjects reported liking a beverage, their brains registered an enhanced response in the ventromedial prefrontal cortex, another region that mediates reward.[20]

When subjects were later shown the brand's label before tasting, however, many changed their preferences. In response to the branding cue, 75 percent of the subjects said that they favored the sample preceded by an image of a Coke can. Montague could tell whether subjects were going to pick Coke or Pepsi by whether two of three regions—the ventral midbrain, the ventral striatum (which includes the nucleus accumbens), and the ventral medial prefrontal cortex—showed enhanced activity in response to one brand over the other. More often than not, Coke prompted the stronger responses. The team interpreted this finding to mean that Coke's success was due to its ability to trigger a frisson of emotionally tinged memories, presumably because of its more effective brand marketing. "There's a huge effect of the Coke label on brain activity related to the control of actions, the dredging up of memories, and things that involve self-image," Montague explained.[21]

The Coke-Pepsi study was a media sensation. *Time, Newsweek,* the British *Guardian, Frontline, PBS,* and other outlets covered it. Soon, the central metaphor for the experiment became the "buy button," invoked by *Newsweek* ("Pushing the Buy Button in the Brain"), *Forbes* ("In Search of the Buy Button"), and the *New York Times* ("If Your Brain has a 'Buy Button', What Pushes It?"). Advertisers also loved the experiment. They embraced it as a dramatic lesson in the role of emotion in determining the power of branding. Some industry insiders credit the experiment with jump-starting the field of neuromarketing.[22]

Can measuring brain function directly predict sales or advertising success better than existing methods? To some degree yes, according to a much-cited 2007 study by neuroscientist Brian Knutson and colleagues. They scanned subjects with fMRI as they viewed pictures of products including a box of Godiva chocolates, a *Sex and the City* DVD, and a smoothie maker. Subjects were allowed to purchase these items with actual money they were given by experimenters, but not until they viewed the item again, this time with an accompanying price tag. After several seconds, subjects pressed a button to indicate whether they wanted to buy it. The researchers found that activation in regions associated with anticipating gain (the nucleus accumbens) correlated with product preference, while activation in regions associated with anticipating loss (the insula) correlated with excessive prices. Further, activation in a region implicated in integrating gains and losses (the mesial prefrontal cortex) correlated with reduced prices. This suggested to the team that activation of distinct brain regions related to anticipation of gain and loss precedes and can be used to predict purchasing decisions. Their accuracy rate of 60 percent was not vastly greater than chance, although it was a little higher than the accuracy of subjects' self-reported preferences for the various items just before they pressed the "purchase" button.[23]

In 2011, neuroscientists Gregory Berns and Sara Moore also attracted media coverage for their work predicting the commercial success of new songs. They asked adolescents to listen to clips of 120 new recordings from unknown artists while they were in an fMRI scanner. Strong responses in the nucleus accumbens, part of the reward pathway, identified about a third of the songs whose albums went on to sell more than 20,000 copies, and weak responses in the nucleus accumbens and orbitofrontal cortex were about 80 percent accurate in predicting tunes that sold fewer than 20,000 copies. Notably, how likable the subjects said they found the songs did not predict sales: The activity within the nucleus accumbens, though, did correlate with the number of units sold. Perhaps, as Berns and Moore suggest, neural markers of specific acoustic or lyric features that predict success could one day be used by composers to reverse-engineer new songs.[24]

NEUROMARKETERS differ from consumer neuroscientists. The former are less interested in how the brain operates during choice making than in what its human proprietor "chooses"—and in how to tempt the brain to "choose" their clients' products. Neuromarketers' services don't come cheap; an average EEG or fMRI marketing study costs around $40,000 to $50,000.[25] Still, there seems to be no shortage of willing clients.

For example, a Coca-Cola marketing team used EEG to help edit an ad for Super Bowl XLII in 2008. After screening several possible commercials for volunteers, the marketers noticed that the viewers appeared to be more "engaged" when the music in one version of the commercial built to a crescendo. The ad team altered its original version of the commercial accordingly. Reportedly, creative teams working with a number of modern big-budget films, including *Avatar*, used EEG measures of viewers' brain responses to different scenes and sequences to help them refine film elements, such as scripts, characters, plots, scenes, effects, and even cast selection. MindSign, a neuromarketing company in San Diego, deployed fMRI to help develop the most engaging movie trailer possible for Warner Brothers' *Harry Potter and the Half-Blood Prince*. The researchers showed film sequences to test audiences to measure their levels of attention and emotional reactions, such as pleasure, fear, boredom, or empathy.[26]

When Pantene, a maker of hair products, wanted to explore women's "overall feelings about their hair," in the words of a lead scientist at Procter and Gamble, it enlisted NeuroFocus. The NeuroFocus analysts recorded electrical signals at the surface of women's heads as they watched a Pantene commercial, creating a millisecond-by-millisecond picture of activity in the brain. According to the brain-wave data, the women became "distracted" at the point in the commercial when a model appeared frustrated as she tried to deal with her unruly hair. Procter and Gamble revised the ad to focus more on the model's hair and less on her facial expression.[27]

But how meaningful are these conclusions? One is tempted to assume that they have value, given that reputable companies use brain-generated information. Yet the lack of transparency surrounding neuromarketers' interpretation of these data opens them to challenge by critics. Columbia University researchers recently reviewed the websites of sixteen neuromarketing firms and found that few described their methodology with enough detail to verify their claims. Almost half of the companies did not even use EEG or fMRI but rather relied on old technologies like skin conductance response or measures of pupil size. Moreover, neuromarketing companies' use of different proprietary formulas for interpreting brain-wave data makes it even more difficult to assess their utility.

NeuroFocus, for example, claims to detect responses along seven dimensions: attention, emotional engagement, memory retention, overall effectiveness, purchase intent, novelty, and awareness. Although there is a large body of research on the relationship between EEG and attention, emotion, and retention of information, NeuroFocus uses a complex proprietary formula to transform those data into measurements that the company states reflect "purchase intent." NeuroFocus also interprets electrical activity over the inferior frontal lobe as reflecting engagement of mirror neurons—cells that are implicated, some experts contend, in human empathy—thereby reflecting a subject's desire to share in the experience of the people depicted in an ad.[28] This is a controversial interpretation because the significance of mirror neurons in humans is not well understood.

Evaluating commercials or movie trailers poses different challenges from those of static formats such as ads in magazines or product designs. For one thing, the neural responses elicited by watching a commercial may not be reliably in sync with what subjects are viewing in real time. This is because brain activation may reflect what a subject anticipates, not what is on the screen at the moment. Also, the density of features in a commercial or movie trailer—the dialogue, music, and images—makes it difficult for analysts to discern the emotional impact of a specific feature.[29]

The caveats don't end there. Neuromarketers can also run afoul of reverse inference, the practice of reasoning backward from regional brain activity to conclude that subjects are thinking certain thoughts or having certain feelings. Imaging results recently prompted Frito-Lay to change its packaging for potato chips from shiny to matte paper. When women viewed the regular, shiny packaging, scans depicted more activity in their anterior cingulate cortex—"an area of the brain associated with feelings of guilt [over eating junk food]," as *Forbes* reported it—than when they viewed a package made of matte beige paper. Yet the anterior cingulate cortex is one of the most promiscuously excitable structures in the brain, participating in the perception of pain, emotional engagement, depression, motivation, error prediction, conflict monitoring, decision making, and more.[30]

Reverse inference also found its way into fMRI analysis of the Super Bowl XL halftime ads in 2006. As subjects watched ads that had been broadcast during the game, neuroscientist Marco Iacoboni scanned their brains. He proclaimed one commercial a "flop" for FedEx, involving a hapless caveman who is fired by his boss because he did not use the carrier to deliver a package. Why? Because when the caveman is subsequently crushed by a dinosaur, the subjects' amygdalae showed enhanced activity. "The scene looks funny, and has been described as funny by lots of people," he said, "but your amygdala still perceives it as threatening." As we know, however, the amygdala does much more than just process fear. Among other things, it mediates response to novelty—and a new Super Bowl ad is nothing if not novel. Even if the scan captured a fearlike response, "fear" experienced within a safe setting can be exhilarating, as any roller-coaster fan can tell you. Thus, when self-report ("this is funny") clashes with what the brain seems to be saying ("I'm scared"), caution is in order. Should FedEx scrap the ad lest it scare potential customers? Of course not. Sure enough, Iacoboni also panned a 2006 Super Bowl ad for GoDaddy.com, a Web-hosting company, because it did not increase activity in brain areas linked to reward (a curious finding in and of itself because the ad features buxom spokesladies). Yet in the real world, the Hooters-esque ad scored a touchdown of sorts, driving more traffic to the advertiser's website than any other website during the game.[31]

"ADVERTISING SEEKS TO PERSUADE and everybody knows it," wrote John E. Calfee in his book *Fear of Persuasion*.[32] Self-interested sellers and skeptical consumers have been enduring features of commerce since the beginning of time. But when audiences fear that they are being manipulated without knowing how—and will thus be incapable of resisting—skepticism can morph into anger and paranoia.

The specter of consumer manipulation memorably roiled the public in 1957 with the publication of Vance Packard's clarion call *The Hidden Persuaders*. The journalist and social critic charged marketers in general, and Ernest Dichter in particular, with undermining the rational autonomy of citizens by manipulating them into buying things they neither wanted nor needed. "Large-scale efforts," Packard wrote, "are being made, often with impressive success, to channel our unthinking habits, our purchasing decisions, and our thought processes. . . . The result is that many of us are being influenced and manipulated, far more than we realize, in the patterns of our everyday lives. The aim is nothing less than to influence the state of our mind and to channel our own behavior as citizens." The *New Yorker* called Packard's book "a brisk, authoritative and frightening report on how such pressure groups as manufacturers, fundraisers and politicians are attempting, with the help of advertising agencies and publicists, to turn the American mind into a kind of catatonic dough that will buy, give or vote at their command." A review of Packard's book in the *Texas Law Review* wondered whether use of subliminal advertising in TV election ads might mean that "Orwell's

1984 is nearer than its title portends"—and whether First Amendment law can cope with "questions raised by Packard regarding recent advances in the science of molding men's minds."[33]

The Hidden Persuaders spent six weeks as the number-one best-selling nonfiction book in the United States. Its thesis resonated during the Cold War era and its attendant fears of communism. Rumors that American POWs had been "brainwashed" during the Korean War were widely believed, inspiring Richard Condon's 1959 novel and later film *The Manchurian Candidate*. Senator Joseph McCarthy fomented fear that Communist spies and sympathizers had infiltrated the federal government and even the U.S. Army. From 1954 to 1960, the Cincinnati Reds baseball team rechristened itself the "Cincinnati Redlegs" to avoid any association with the Communist contagion.[34] The famous 1956 science-fiction movie *Invasion of the Body Snatchers*—in which large alien pods produced replacement human beings—has been interpreted as a political allegory of fears that Communist ideology would eradicate individualism and impose a soulless conformity on the country.

Against this backdrop, James Vicary, a Manhattan marketing executive and founder of the Subliminal Projection Company, claimed that he had devised a technique he called "subliminal advertising" to change buyer behavior. ("Subliminal" refers to an image or sound, usually briefly presented, that fails to register in consciousness.) Subliminal persuasion, Vicary's elusive goal, should not be confused with the well-documented phenomenon of subliminal perception, the well-documented capacity to perceive input without being aware of doing so.) Five months after Packard published *The Hidden Persuaders*, Vicary held a press conference to announce the success of his "invisible commercial." During that summer, Vicary told attendees, he had flashed the messages "Hungry? Eat Popcorn" and "Drink Coca-Cola" for an imperceptible 1/3,000th of a second during showings of the movie *Picnic* at a theater in Fort Lee, New Jersey. Vicary's results were impressive, to say the least. Six weeks of subliminal exposure, Vicary claimed, spiked sales of popcorn and Coca-Cola at the theater by 18 and 58 percent, respectively.[35]

Vicary's subliminal mandate to drink Coke and eat popcorn sparked a public furor. "Welcome to 1984," wrote Norman Cousins, the legendary editor of *Saturday Review*. "If the device is successful for putting over popcorn, why not politicians or anything else?" Polls revealed widespread condemnation by the public, and Congress called for the Federal Communications Commission (FCC) to regulate subliminal advertising. Meanwhile, the National Association of Radio and Television Broadcasters asked its member stations to refrain from using subliminals pending review.[36]

From the start, the Advertising Research Foundation and most research psychologists were skeptical of Vicary's claims. They insisted that Vicary provide his data or repeat his demonstration. In January 1958, Vicary traveled to the nation's capital to demonstrate his "technique" before several members of Congress and the FCC chairman. In a Washington television studio, Vicary showed them a few minutes of a movie with split-second "Eat Popcorn" messages inserted in the film, but could not spur any desire for popcorn. One lawmaker, however, reportedly quipped that the film made him want a hot dog. The following month, the Canadian Broadcasting Corporation announced that it had attempted its own subliminal persuasion experiment, flashing a hidden message—"Phone now"—throughout a half-hour show. Of five hundred viewers surveyed, only one reported an urge to make a phone call. Many viewers said that the broadcast made them feel hungry or thirsty.[37]

Finally, in 1962, Vicary told *Advertising Age* that his "experiment" consisted of a handful of data he sloppily collected and merged with falsified evidence. The main objective for his press conference, Vicary admitted, was to attract publicity for his consulting company. Psychologist Raymond A. Bauer would not have been surprised. Writing in the *Harvard Business Review* in 1958, he observed that the "specter of 'manipulation' and 'hidden persuasion' has stalked all the lands that man never inhabited." Subsequently, numerous well-designed studies have shown that subliminal messages do not readily manipulate individuals or groups to change their purchasing behaviors. Admittedly, some relatively recent laboratory evidence raises the possibility that subliminal messages may at times affect our motivations; for example, in one study, participants exposed subliminally to photos of Coke cans and the word "thirsty" later reported being thirstier than participants not exposed to the word. Yet it is not at all clear that these findings translate into real-world purchasing decisions, let alone that they can shape preferences for specific brands.[38]

Public reaction to neuromarketing has been far less dramatic than it was to Packard's book and Vicary's revelations. Yet latent anxiety still surrounds the prospect that we are being manipulated in new ways and that the potential for control will only grow. Neuromarketing is already in the crosshairs of vigilant consumer-protection groups. In 2003, Commercial Alert, a nonprofit organization co-founded by Ralph Nader, appealed to the Department of Health and Human Services about research conducted by the BrightHouse Institute for Thought Sciences and Emory University, asking, "What exactly will stop Emory's neuromarketing research from being sold to corporate clients to push the 'buy button' for more tobacco, alcohol, junk food, violence, gambling and other addictive or destructive behaviors?" The following year, the group unsuccessfully urged the Senate Commerce Committee to undertake a federal investigation of BrightHouse.[39]

In 2011, a consortium of consumer-protection groups filed a complaint with the Federal Trade Commission against Frito-Lay for allegedly using neuromarketing "designed to trigger subconscious, emotional arousal" in order to promote high-fat snack food to teens. Richard Glen Boire of the Center for Cognitive Liberty and Ethics in Davis, California, suggests that use of neuromarketing techniques be disclosed by companies: "If the technology becomes very effective,

we might see some companies adopt a No-Neuromarketing position, much as we see on some personal products a note that they are
not tested on animals."[40] Boire's remark conjures a little symbol in the
corner of a package label—a brain with a slash through it, perhaps—
indicating that no brains were examined in the making of this
product.

Legally speaking, there is no federal law against the use of subliminal messages in advertisements. The FCC may revoke a company's broadcast license if the use of "subliminal techniques" is proved,
irrespective of its effectiveness. And state and federal judges and
scholars have generally taken the position that the First Amendment
should not protect subliminal advertising and have often ruled
against its use. According to legal scholar Marc J. Blitz, when a communication contains messages or stimuli that are designed to influence our thinking without our being aware of them, the logic for
First Amendment protection disappears. That is, if people are unaware of information that is influencing them, they cannot subject it
to analysis. And if they cannot analyze a message to determine its
truth through debate and dialogue—because they are unaware of
the message in the first place—the constitutional protection would
not apply. Granted, advertising, like all communication, can affect
us in ways of which we are not aware. Does neuromarketing—should
it someday be demonstrated to be unequivocally effective—represent
a deeper violation of autonomy? The ultimate question, Blitz says,
is whether neuromarketing influences consumers' behavior strongly
enough to threaten their autonomy.[41]

We do not believe that Manchurian customers will be marching
down department store aisles any time soon, if ever. Consumers
aren't disembodied brains milling about the Mall of America. They
juggle their pocketbooks and contemplate other items they have recently bought. Purchasing is a social activity, and people are social
creatures, gauging the foreseeable reaction from a spouse ("You
bought *what*!?") and often soliciting advice from family, friends, or

experts before buying. Indeed, influences on buyers are ubiquitous in the environment as well. Shoppers' moods, for example, influence buying behavior. So can the ambient pace of music in a store. Higher levels of arousal seem to prompt people to process information more shallowly—that is, to rely more heavily on cognitive biases and shortcuts—thus making them more easily swayed by celebrity product endorsers, the catchiness of an ad, or other superficial yet attractive elements. Mentally drained individuals may be more likely to choose superficially entertaining, lowbrow movies over more cerebral, slower-paced films, which implies that the nature of a television program itself (say, the postapocalyptic drama *The Walking Dead* versus the family sitcom *The Big Bang Theory*) can influence how people perceive commercials.[42]

In the end, a cacophony of influences impinge on us at once, some cancelling out others, some combining in novel ways, some emanating from within us, some from the external environment, and still others generated by advertisers. Our implicit unconscious processes and overt conscious capacities come together to guide us.

So is neuromarketing "hidden persuasion or junk science," as *Advertising Age* asked in 2007? It is neither. There are limits to influencing human behavior, in general, and there is no specific evidence that neuromarketers can manipulate information they glean from our brains to turn us into passive, unconscious consumers of things we don't need. As esteemed market researcher Andrew S. C. Ehrenberg wrote in 1982, "Advertising is in an odd position. Its extreme protagonists claim it has extraordinary powers . . . and its severest critics believe them." Three decades later, his observation is just as true. Nor is it fair to allow buyologists' exaggerated claims to taint all neuromarketing and dismiss it out of hand as junk science. For one thing, its premise is sound: namely, that people are drawn to certain products and disposed to purchase them for motives to which they are often not privy. It may turn out that neuromarketing is best suited for generating and testing early hypotheses about the optimal way to grab viewers' attention and engage them emotionally. For example, if moments in the initial versions of a commercial or movie clip arouse only very weak responses, the team may want to go back to the drawing board.[43]

At bottom, however, the predictive value of neural information will take on real marketplace significance only if it outperforms what people say they will buy or what they say they like about a product. If this is already happening—and given the paucity of available, replicated evidence, we are skeptical—neuromarketers are not sharing their in-house proprietary data and their methods. What's more, the "neuro" part of marketing needs to be worth it compared with conventional methods. "If I can spend $1,000 to do a traditional market study that gets me 80 percent of what a $24,000 fMRI study does, then the return on my neuromarketing investment is not great," says neuroscientist Craig Bennett, of dead salmon fame.[44]

Nonetheless, the burden falls on neuromarketing to prove itself. In 2010, the Advertising Research Foundation began a long-term project to develop neuromarketing guidelines. After a review of methods used by a number of neuromarketing firms, the foundation concluded that "the complexity of the science underlying these methods makes it difficult to assess their validity." The project's reviewers noted what they saw as neuromarketers' too-frequent exaggeration of what their tests could deliver and asserted that "documentation of methods, research protocols, and clarity about what was done are essential," given the complexities involved.[45]

For now, the basics of advertising remain intact. Effective advertising must be seen, read, believed, remembered, and acted on—much as pioneering market researcher Daniel Starch concluded in the 1920s. Marketers still evaluate promotional campaigns and products according to traditional constructs: Are viewers paying attention to an ad, do they like it, can they recognize and recall the product, do they identify with the brand image, and do they intend to purchase?

Marketers continue to rely heavily on surveys, market tests of product samples, one-on-one interviews with consumers, and, yes, old-fashioned focus groups. Whether neuromarketing will flourish, burn out, or flicker on the periphery of the advertising world remains to be seen. Right now, the promises are bright and shiny, but behind the scenes, the fallacies and pitfalls of overhyped neuroscience give the "[corporate] buyer beware" truism a new twist.[46]

In the next chapter, we continue the theme of the biology of desire and decision making, this time from the angle of addiction to alcohol and drugs. Can an examination of the brains of people with addiction reveal insights that researchers and doctors can use to help them in treatment and recovery? As we will see, neuroscientific findings are fascinating, but the overriding emphasis on the brain—the now-dominant approach to the study of addiction—is too narrow a perspective.

3

ADDICTION AND THE
BRAIN-DISEASE FALLACY

IN 1970, HIGH-GRADE HEROIN and opium flooded Southeast Asia. Military physicians in Vietnam estimated that nearly half of all U.S. Army enlisted men serving there had tried opium or heroin, and between 10 and 25 percent of them were addicted. Deaths from overdoses soared. In May 1971, the crisis reached the front page of the *New York Times*: "G.I. Heroin Addiction Epidemic in Vietnam." Fearful that the newly discharged veterans would join the ranks of junkies already bedeviling inner cities, President Richard Nixon commanded the military to begin drug testing. No one could board a plane home until he had passed a urine test. Those who failed could attend an army-sponsored detoxification program.[1]

Operation Golden Flow, as the military called it, succeeded. As word of the new directive spread, most GIs stopped using narcotics. Almost all the soldiers who were detained passed the test on their second try. Once they were home, heroin lost its appeal. Opiates may have helped them endure a war's alternating bouts of boredom and terror, but stateside, civilian life took precedence. The sordid drug culture, the high price of heroin, and fears of arrest discouraged use, veterans told Lee Robins, the Washington

University sociologist who evaluated the testing program from
1972 to 1974.[2]

Robins's findings were startling. Only 5 percent of the men who
became addicted in Vietnam relapsed within ten months after return, and just 12 percent relapsed briefly within three years. "This
surprising rate of recovery even when re-exposed to narcotic drugs,"
wrote Robins, "ran counter to the conventional wisdom that heroin
is a drug which causes addicts to suffer intolerable craving that rapidly leads to re-addiction if re-exposed to the drug." Scholars hailed
the results as "revolutionary" and "path-breaking." The fact that
addicts could quit heroin and remain drug free overturned the belief
that "once an addict, always an addict."[3]

Unfortunately, that lesson has faded into the past. By the mid-
1990s, the truism "once an addict, always an addict" was back, re-
packaged with a new neurocentric twist: "Addiction is a chronic and
relapsing brain disease." It was promoted tirelessly by psychologist
Alan I. Leshner, then the director of the National Institute on Drug
Abuse (NIDA), the nation's premier addiction research body and
part of the National Institutes of Health, and is now the dominant
view of addiction in the field. The brain-disease model is a staple of
medical school education and drug counselor training and even appears in the antidrug lectures given to high-school students. Rehab
patients learn that they have a chronic brain disease. And the American Society of Addiction Medicine, the largest professional group
of physicians specializing in drug problems, calls addiction "a primary, chronic disease of brain reward, motivation, memory and related circuitry." Drug czars under Presidents Bill Clinton, George W.
Bush, and Barack Obama have all endorsed the brain-disease framework at one time or another. From being featured in a major documentary on HBO, on talk shows and *Law and Order*, and on the
covers of *Time* and *Newsweek*, the brain-disease model has become
dogma—and like all articles of faith, it is typically believed without
question.[4]

That may be good public relations, but it is bad public education. We also argue that it is fundamentally bad science. The brain-disease model of addiction is not a trivial rebranding of an age-old human problem. It plays to the assumption that if biological roots can be identified, then a person has a "disease." And being afflicted means that the person cannot choose, control his or her life, or be held accountable. Now introduce brain imaging, which seems to serve up visual proof that addiction is a brain disease. But neurobiology is not destiny: The disruptions in neural mechanisms associated with addiction do constrain a person's capacity for choice, but they do not destroy it. What's more, training the spotlight too intently on the workings of the addicted brain leaves the addicted person in the shadows, distracting clinicians, policy makers, and sometimes patients themselves from other powerful psychological and environmental forces that exert strong influence on them.

For over three centuries in the United States, physicians, legal scholars, politicians, and the public have debated the nature of addiction: Is it a defect of the will or of the body? A moral or a medical problem? Such polarization should by now have exhausted itself.[5] After all, mountains of evidence attest to the fact that addiction entails both biological alterations in the brain and in deficits personal agency. But given what is at stake in these debates—namely, our deep cultural beliefs about self-control and personal responsibility paired with concerns about what society owes to addicts and what it can expect of them—we must be very careful not to ascribe too much influence to the addict's brain.

WHAT exactly makes addiction a brain disease? "That addiction is tied to changes in brain structure and function is what makes it, fundamentally, a brain disease," Leshner wrote in a now-landmark article in *Science* in 1997. But that can't be right. Every experience changes the brain—from learning a new language to navigating a new city. It is certainly true that not all brain changes are equal;

learning French is not the same as acquiring a crack habit. In addiction, intense activation of certain systems in the brain makes it difficult for users to quit. Genetic factors influence the intensity and quality of the subjective effect of the drug, as well as the potency of craving and the severity of withdrawal symptoms.[6]

The process of addiction unfolds partly through the action of dopamine, one of the brain's primary neurotransmitters. Normally, dopamine surges in the so-called reward pathway, or circuit, in the presence of food, sex, and other stimuli central to survival. Dopamine enhancement serves as a "learning signal" that prompts us to repeat eating, mating, and other pleasures. Over time, drugs come to mimic these natural stimuli. With every puff of a Marlboro, injection of heroin, or swig of Jim Beam, the learning signal in the reward pathway is strengthened, and in vulnerable users, these substances assume incentive properties reminiscent of food and sex.

"Salience" is the term that neuroscientists often use to describe the pull of substances on the addicted—it's more of a sense of wanting, even needing, than liking. The development of salience has been traced to the nerve pathways that mediate the experience as they emerge from the underside of the brain, in an area called the ventral tegmentum, and sweep out to regions such as the nucleus accumbens, hippocampus, and prefrontal cortex, which are associated with reward, motivation, memory, judgment, inhibition, and planning.

Other nerve fibers travel from the prefrontal cortex, a region involved in judgment and inhibition, to parts of the brain that control behavior. As one psychiatrist put it memorably, "The war on drugs is a war between the hijacked reward pathways that push the person to want to use, and the frontal lobes, which try to keep the beast at bay." Note the word "hijacked." As shorthand for the usurpation of brain circuitry during the addiction process, it is a reasonable metaphor. In the hands of brain-disease purists, though, "hijacking" has come to denote an all-or-nothing process, likened to a "switch in the brain" that, once flipped, affords no retreat for the addict. "It may start with the voluntary act of taking drugs," Leshner said, "but once you've got [addiction], you can't just tell the addict, 'Stop,' any more than you can tell the smoker 'Don't have emphysema.'"[7]

The reward circuit is also intimately involved in "cue-induced" craving. Such craving is a special species of desire that manifests itself in a sudden, intrusive urge to use brought on by cues associated with use. The mere clink of a whiskey bottle, a whiff of cigarette smoke, or a glimpse of an old drug buddy on the corner can set off an unbidden rush of yearning, fueled by dopamine surges. For the addict who is trying to quit, this is a tense feeling, not pleasurable at all. Because the rush of desire seems to come out of the blue, users may feel blindsided, helpless, and confused.[8]

In a very impressive display of brain technology, scientists have used PET and fMRI scans to observe the neural correlates of craving. In a typical demonstration, addicts watch videos of people handling a crack pipe or needle, causing their prefrontal cortices, amyadalae, and other structures to bloom with activity. (Videos of neutral content, such as landscapes, induce no such response.)[9] Even in users who quit several months previously, neuronal alterations may persist, leaving them vulnerable to sudden, strong urges to use. The familiar late-1980s slogan "This is your brain on drugs" is still with us, but now with the brain itself substituting for the fried egg.

But that egg is not always sizzling. There is a surprising amount of lucid time in the daily life of addicts. In their classic 1969 study "Taking Care of Business: The Heroin User's Life on the Street," criminologists Edward Preble and John J. Casey found that addicts spend only a small fraction of their days getting high. Most of their time is spent either working or hustling. The same is true for many cocaine addicts.[10] We tend to think of them, at their worst, frantically gouging their skin with needles, jamming a new rock into a pipe every fifteen minutes, or inhaling lines of powder. In the grip of such hunger, an addict cannot be expected blithely to get up and walk away.

These tumultuous states—with neuronal function severely disrupted—are the closest drug use comes to being beyond the user's restraint. But in the days between binges, cocaine users worry about a host of everyday matters: Should I find a different job? Enroll my kid in a better school? Kick that freeloading cousin off the couch for good? Attend a Narcotics Anonymous meeting, enter treatment, or register at a public clinic? It is during these stretches of relative calm that many addicts could make the decision to get help or quit on their own—and many of them do. But the decision to quit can be long in coming, far too long for those who destroy their health, families, or careers in the meantime.

The paradox at the heart of addiction is this: How can the capacity for choice coexist with self-destructiveness? "I've never come across a single person that was addicted that wanted to be addicted," says neuroscientist Nora Volkow, who succeeded Leshner as director of NIDA in 2003. Exactly. How many of us have ever come across a heavy person who wanted to be overweight? Many undesirable outcomes in life arrive incrementally. "We can imagine an addict choosing to get high each day, though not choosing to be an addict," says psychologist Gene Heyman. "Yet choosing to get high each day makes one an addict."[11]

Let's follow a typical trajectory to see how this dynamic plays out. In the early phase of addiction, drugs or alcohol become ever more appealing, while once-rewarding activities, such as relationships, work, or family, decline in value. The attraction of the drug starts to fade as consequences accrue—spending too much money, disappointing loved ones, attracting suspicion at work—but the drug still retains its allure because it blunts psychic pain, suppresses withdrawal symptoms, and douses intense craving.[12] Addicts find themselves torn between the reasons to use and reasons not to.

Sometimes a spasm of self-reproach or a flash of self-awareness tips the balance toward quitting. William S. Burroughs, an American novelist and heroin addict, calls this the "naked lunch" experience, "a frozen moment when everyone sees what is on the end of every fork." Christopher Kennedy Lawford, himself in recovery from drugs and alcohol, edited a 2009 collection of essays called *Moments of Clarity* in which the actor Alec Baldwin, singer Judy Collins, and others recount the events that spurred their recoveries. Some quit on their own; others got professional help. A theme in each of their stories is a jolt to self-image: "This is not who I am, not who I want to be." One recovered alcoholic describes the process: "You tear yourself apart, examine each individual piece, toss out the useless, rehabilitate the useful, and put your moral self back together again."[13] These are not the sentiments of people in helpless thrall to their diseased brains. Nor are these sentiments the luxury of memoirists. Patients have described similar experiences to us: "My God, I almost robbed someone!" "What kind of mother am I?" or "I swore I would never switch to the needle."

And it turns out that quitting is the rule, not the exception—a fact worth acknowledging, given that the official NIDA formulation is that "addiction is a *chronic and relapsing* [italics added] brain disease." The Epidemiologic Catchment Area Study, done in the early 1980s, surveyed 19,000 people. Among those who had become dependent on drugs by age twenty-four, more than half later reported not a single drug-related symptom. By age thirty-seven, roughly 75 percent reported no drug symptom. The National Comorbidity Survey, conducted between 1990 and 1992 and again between 2001 and 2003, and the National Epidemiologic Survey on Alcohol and Related Conditions, conducted between 2001 and 2002 with more than 43,000 subjects, found that 77 and 86 percent of people who said they had once been addicted to drugs or alcohol reported no substance problems during the year before the survey.[14]

By comparison, people who were addicted within the year before the survey were more likely to have concurrent psychiatric disorders. Additionally, NIDA estimates that relapse rates of treated drug-addicted patients run from 40 to 60 percent.[15] In other words, they are not representative of the universe of addicts. They are the hard cases—the chronic and relapsing patients. Yet these patients often make the biggest impressions on clinicians and shape their views of addiction, if only because clinicians are especially likely to encounter them.

Researchers and medical professionals err in generalizing from the sickest subset of people to the overall population of patients. This caveat applies across the medical spectrum. Just as the clinician wrongly assumes that all addicts must be like the recalcitrant ones who keep stumbling through the clinic doors, psychiatrists sometimes view people with schizophrenia as doomed to a life of dysfunction on the basis of their frequent encounters with those whose delusions and hallucinations don't improve with treatment. The error of extrapolating liberally from these subsets of difficult patients is so common that statisticians Patricia and Jacob Cohen gave it a name: the "clinician's illusion."[16]

ADVOCATES of the brain-disease paradigm have good intentions. By placing addiction on an equal medical footing with more conventional brain disorders, such as Alzheimer's and Parkinson's, they want to create an image of addicts as victims of their own wayward neurochemistry. They hope that this portrayal will inspire insurance companies to expand coverage for addiction and politicians to allocate more funding for treatment. And in the hands of Alan Leshner, the model has had real political utility. Before he was NIDA director, Leshner served as acting director of the National Institute of Mental Health. There, he saw how brain-disease "branding" could prompt Congress to act. "Mental health advocates started referring to schizophrenia as a 'brain disease' and showing brain scans to members of Congress to get them to increase funding for research. It really worked," he said.[17]

Many experts credit the brain-disease narrative with enhancing the profile of their field. The late Bob Schuster, head of NIDA from

1986 to 1991, admitted that although he did not think of addiction as a disease, he was "happy for it to be conceptualized that way for pragmatic reasons . . . for selling it to Congress." For decades, addiction research had been a low-status field, disparaged by other researchers as a soft science that studied drunks and junkies. Now the field of neuroscience was taking greater notice. "People recognize that certain decision makers and others are very impressed with molecular biology," said Robert L. Balster, director of the Institute for Drug and Alcohol Studies at Virginia Commonwealth University.[18]

Psychiatrist Jerome Jaffe, an eminent figure in the field and the first White House adviser on drugs (the precursor of the "drug czar"), sees the adoption of the brain-disease model as both a tactical triumph and a scientific setback. "It was a useful way for particular agencies to convince Congress to raise the budgets, [and] it has been very successful," he said. Indeed, neuroimaging, neurobiological research, and medication development consume over half of the NIDA research budget. In light of the agency's reach—it funds almost all substance-abuse research in the United States—it sets the national agenda regarding which research gets funded and therefore the nature of the data produced and the kinds of topics that investigators propose. But Jaffe argues that the brain-disease paradigm presents "a Faustian bargain—the price that one pays is that you don't see all the other factors that interact [in addiction]."[19]

The proponents of the brain-disease concept are also trying to dispel the stigma surrounding addiction by rehabilitating addicts' poor public image: They are not undisciplined deadbeats but just people struggling with an ailment. This approach had its roots in the world of mental health advocacy. Until the early 1980s, plenty of people blamed parents for their children's serious mental problems. Then advocates began to publicize neuroscientific discoveries, demonstrating, for example, that schizophrenia is associated with abnormalities of brain structure and function. In this effort, brain imaging has served sufferers well, helping legitimize their symptoms by representing visually the illness in their brains.[20] The idea, of course, was that these benefits would extend to addicts. But it turns out that it's harder to destigmatize addiction.

For all its benign aspirations, there are numerous problems with the brain-disease model. On its face, it implies that the brain is the most important and useful level of analysis for understanding and treating addiction. Sometimes the model even equates addiction with a neurological illness, plain and simple.[21] Such neurocentrism has clinical consequences, downplaying the underlying psychological and social reasons that drive drug use.

Recovery is a project of the heart and mind. The person, not his or her autonomous brain, is the agent of recovery. Notably, Alcoholics Anonymous, the institution perhaps most responsible for popularizing the idea that addiction is a disease, employs the term as a metaphor for loss of control. Its founders in the 1930s were leery of using the word "disease" because they thought that it discounted the profound importance of personal growth and the cultivation of honesty and humility in achieving sobriety.[22]

The brain-disease narrative misappropriates language better used to describe such conditions as multiple sclerosis or schizophrenia—afflictions of the brain that are neither brought on by the sufferer nor modifiable by the desire to be well. It offers false hope that an addict's condition is completely amenable to a medical cure (much as pneumonia is to antibiotics). Finally, as we'll see, it threatens to obscure the vast role of personal agency in perpetuating the cycle of use and relapse.

Addicts embarking on recovery often need to find new clean and sober friends, travel new routes home from work to avoid passing near their dealer's street, or deposit their paycheck directly into a spouse's account to keep from squandering money on drugs. A teacher trying to quit cocaine switched from using a chalkboard—the powdery chalk was too similar to cocaine—and had a whiteboard installed instead. An investment banker who loved injecting speedballs—a cocktail of cocaine and heroin in the same syringe—made himself wear long-sleeved shirts to prevent glimpses of his bare and inviting arms.[23] Former smokers who want to quit need to make many fine adjustments, from not lingering at the table after meals to ridding their homes of the ever-present smell of smoke, removing car light-ers, and so on.

Thomas Schelling, a 2005 Nobel laureate in economics, refers to these purposeful practices as self-binding. The great self-binder of myth was Odysseus. To keep himself from heeding the overpower-ing song of the sea sirens—the half-woman, half-bird creatures whose beautiful voices lured sailors to their deaths—Odysseus instructed his men to tie him to the mast of his ship. The famous Romantic En-glish poet Samuel Taylor Coleridge, an opium addict, is said to have hired men to prevent him from entering a pharmacy to purchase opium. Today, one can hire a firm that will provide binding services. It imposes surprise urine tests on the client, collects evidence of at-tendance at AA meetings or treatment sessions, and sends a monthly status report (with the good or bad news) to another person, such as a parent, spouse, or boss.[24]

Some addicts devise their own self-binding strategies. Others need the help of therapists, who teach them to identify and anticipate cues that trigger craving. Beyond the classic triad of people, places, and things, they come to realize that internal states, such as stress, bad moods, and boredom, can prompt drug urges.

Managing craving matters mightily in recovery, but it usually is not enough. Another very important truth is that an addict uses drugs or alcohol because they serve a purpose. Caroline Knapp, in her powerful 1996 memoir *Drinking: A Love Story*, recounted why she spent two decades of her life as an alcoholic: "You drank to drown out fear, to dilute anxiety and doubt and self-loathing and painful memories."[25] Knapp doesn't describe an urge to drink so much as a need to drink. She was not manipulated by an alien desire but by something woven into her being. To say that Knapp's problem was merely the effect of heavy drinking on her brain is to miss the true threat to her well-being: the brilliant but tormented Knapp herself.

Heroin and speed helped screenwriter Jerry Stahl, author of *Permanent Midnight*, attain "the soothing hiss of oblivion." But when the drugs wore off, his vulnerabilities throbbed like a fresh surgical incision. In surveying his life, Stahl wrote, "Everything, bad or good, boil[ed] back to the decade on the needle, and the years before that imbibing everything from cocaine to Romilar, pot to percs, LSD to liquid meth and a pharmacy in between: a lifetime spent altering the single niggling fact that to be alive means being conscious."[26]

Or take Lisa, a thirty-seven-year-old woman featured in an HBO documentary on addiction. When we meet her, Lisa is living in a run-down hotel room in Toronto and working as a prostitute. She sits on the bed and talks with the filmmaker behind the camera. Flipping her shiny brown hair and inspecting her well-kept nails, Lisa is animated as she boasts about how much she makes selling sex, how much she spends on cocaine, and the longed-for "oblivion" that drugs help her attain. When Lisa was filmed, she was healthy and engaging; she looked and talked like someone who had recently been abstinent but was back in the early stages of her next downward spiral. She had no interest in stopping things at this point. "Right now, I am in no position to go into recovery. [This way of life] is working for me. . . . I have money, drugs, business. I'm O.K." To say that Lisa's problem is the effect of cocaine on her brain is to miss the true threat to her well-being: Lisa herself. "I always use for a reason. It's repressing what needs to be repressed," she says.[27]

These stories highlight one of the shortcomings of the neurocentric view of addiction. This perspective ignores the fact that many people are drawn to drugs because the substances temporarily quell their pain: persistent self-loathing, anxiety, alienation, deep-seated intolerance of stress or boredom, and pervasive loneliness. The brain-disease model is of little use here because it does not accommodate the emotional logic that triggers and sustains addiction.[28]

IN December 1966, Leroy Powell of Austin, Texas, was convicted of public intoxication and fined $20 in a municipal court. Powell appealed the conviction to county court, where his lawyer argued that he suffered from "the disease of chronic alcoholism." Powell's public display of inebriation therefore was "not of his own volition," and the fine constituted cruel and unusual punishment. A psychiatrist concurred, testifying that Powell was "powerless not to drink."[29]

Then Powell took the stand. On the morning of his trial, he had a drink at 8 a.m. that his lawyer gave to him, presumably to stave off morning tremors. Here is an excerpt from the cross-examination:

> *Q:* You took that one [drink] at eight o'clock [a.m.] because you wanted to drink?
>
> *A:* Yes, sir.
>
> *Q:* And you knew that if you drank it, you could keep on drinking and get drunk?
>
> *A:* Well, I was supposed to be here on trial, and I didn't take but that one drink.
>
> *Q:* You knew you had to be here this afternoon, but this morning you took one drink and then you knew that you couldn't afford to drink anymore and come to court; is that right?
>
> *A:* Yes, sir, that's right.
>
> *Q:* Because you knew what you would do if you kept drinking that you would finally pass out or be picked up?
>
> *A:* Yes, sir.
>
> *Q:* And you didn't want that to happen to you today?
>
> *A:* No, sir.
>
> *Q:* Not today?
>
> *A:* No, sir.
>
> *Q:* So you only had one drink today?
>
> *A:* Yes, sir.[30]

The judge let stand Powell's conviction for public intoxication. A second appeal followed, this time to the U.S. Supreme Court. It, too, affirmed the constitutionality of punishment for public intoxication. "We are unable to conclude," said the court, "that chronic alcoholics in general, and Leroy Powell in particular, suffer from such an irresistible compulsion to drink and to get drunk in public that they are utterly unable to control their performance."[31]

For people like Powell who are not otherwise motivated to quit, consequences can play a powerful role in modifying behavior. Powell took only a single drink on the morning of his trial because of foreseeable and meaningful consequences. Far from being unusual, his ability to curtail his drinking accords with a wealth of studies showing that people addicted to all kinds of drugs—nicotine, alcohol, cocaine, heroin, methamphetamines—can change in response to rewards or sanctions.[32] Powell had surely experienced many alcohol-induced brain changes, but they did not keep him from making a choice that morning.

If Powell came before a judge today, his lawyer might well introduce a scan of his brain "craving" alcohol as evidence of his helplessness. If so, the judge would be wise to reject the scan as proof. After all, a judge, or anyone, can ponder scans of "addicted" brains all day, but he or she would never consider someone an addict unless that person behaves like one.[33]

Consider the following fMRI experiment by researchers at Yale and Columbia. They found that the brains of smokers reporting a strong desire to smoke displayed enhanced activation of reward circuitry, as would be expected. But they also showed that subjects could reduce craving by considering the long-term consequences of smoking, such as cancer or emphysema, while observing videos depicting people smoking. When subjects did so, their brains displayed enhanced activity in areas of the prefrontal cortex associated with focusing, shifting attention, and controlling emotions. Simultaneously, activity in regions associated with reward, such as the ventral striatum, decreased.[34]

Investigators at NIDA observed the same pattern when they asked cocaine users to inhibit their craving in response to cues. Subjects underwent PET scanning as they watched a video of people preparing drug paraphernalia and smoking crack cocaine. When researchers instructed the addicts to control their responses to the video, they observed inhibition of brain regions normally implicated in drug craving. When not deliberately suppressing their cravings, the addicts reported feeling their typical desire to use, and the PET scans, accordingly, revealed enhanced activation in brain regions that mediate craving.[35]

These powerful findings illuminate the capacity for self-control in addicts. They also underscore the idea that addicts persist not because of an inability to control the desire to use but from a failure of motivation. Granted, summoning sustained motivation can be a great challenge: It takes a lot of energy and vigilance to resist craving, especially urges that ambush the addict unexpectedly. Studies on the regulation of craving also help distinguish behavior that people do not control from behavior that they cannot control. Imagine, by way of contrast, promising a reward to people with Alzheimer's if they can keep their dementia from worsening. That would be both pointless and cruel because the kinds of brain changes intrinsic to dementia leave the sufferer resistant to rewards or penalties.

What Powell's case showed was that even though he sustained brain changes, those changes did not prevent his behavior from being shaped by consequences. Contingency management—the technical term for the practice of adjusting consequences, including incentives—often succeeds with people who face serious losses, such as their livelihood, professional identity, or reputation. When addicted physicians come under the surveillance of their state medical boards and are subject to random urine testing, unannounced workplace visits, and frequent employer evaluations, they fare well: 70 to 90 percent are employed with their licenses intact five years later.[36] Likewise, scores of clinical trials show that addicts who know they will receive a reward, such as cash, gift certificates, or services, are nearly two to three times as likely to submit drug-free urine samples as addicts not offered rewards.[37]

Unfortunately, treatment programs are rarely in a position to offer cash or costly rewards. But the criminal justice system has an ample supply of incentives at its disposal and has been using such leverage for years. One of the most promising demonstrations of contingency management comes from Honolulu in the form of Project HOPE, Hawaii's Opportunity Probation with Enforcement.[38]

Project HOPE includes frequent random drug testing of offenders on probation. Those who test positive are subject to immediate and brief incarceration. Sanctions are fair and transparent: All offenders are treated equally, and everyone knows what will happen in case of an infraction. The judges express a heartfelt faith in offenders' ability to succeed. These basic elements of HOPE's contingency administration—swiftness, sureness, transparency, and fairness combined with expectation for achievement—are a potent prescription for behavior change in just about anyone.[39]

Indeed, after one year of enrollment in Project HOPE, participants fared considerably better than probationers in a group who served as a comparison. They were 55 percent less likely to be arrested for a new crime and 53 percent less likely to have had their probation revoked. These results are even more impressive in light of the participants' criminal histories and their heavy, chronic exposure to methamphetamine, which can impair aspects of cognitive function.[40]

These findings join a vast body of experimental data attesting to the power of incentives to override the lure of drugs. Yet because the facts contradict the idea that addiction is analogous to Alzheimer's disease, some HOPE personnel objected to incentives, arguing that addicts couldn't be accountable for their behavior. Likewise, when researchers asked NIDA to consider reviewing HOPE in its formative years, the agency declined on the grounds that methamphetamine addicts are not capable of responding to incentives alone.[41]

THE brain-disease model leads us down a narrow clinical path. Because it states that addiction is a "chronic and relapsing" condition, it diverts attention from promising behavioral therapies that challenge the inevitability of relapse by holding patients accountable for their choices. At the same time, because the model implies that addicts cannot stop using drugs until their brain chemistry returns to normal, it overemphasizes the value of brain-level solutions, such as pharmaceutical intervention. In 1997, Leshner ranked the search for a medication to treat methamphetamine addiction as a "top priority." A decade later, Volkow predicted, "We will be treating addiction as a disease [by 2018], and that means with medicine."[42]

The search for a magic bullet is folly—and even NIDA has given up hope of finding a wonder drug—but the brain-disease narrative continues to inspire unrealistic goals. When British pop star Amy Winehouse succumbed to her high-profile alcoholism in the summer of 2011, a *Psychology Today* columnist asked, "Could neuroscience have helped Amy Winehouse?" The author answered in the affirmative, suggesting a dopamine-altering medication of the future because addiction "may be a brain problem that science can eventually solve." Neuroscientist David Eagleman goes even further, asserting that "addiction can be reasonably viewed as a neurological problem that allows for medical solutions, just as pneumonia can be viewed as a lung problem."[43] But the analogy doesn't hold up. Changing a behavior like addiction requires addicts to work hard to change their patterns of thought and behavior. In contrast, antibiotic cures for pneumonia work even if the patient is in a coma.

The hope of a medical treatment is the logical outgrowth of placing the brain at the center of the addictive process. Overall, success to date has been genuine but modest. When motivated patients take medications—especially patients already armed with relapse-prevention strategies and the support of family and friends—they can sometimes vault into sustained recovery. Methadone, a long-acting synthetic opiate taken once a day to prevent opiate withdrawal, has played a major role in treating addiction to heroin and painkillers since the 1960s. Still, to their counselors' chagrin, up to half the patients in methadone clinics also fortify themselves with heroin, cocaine, or Valium-like tranquilizers called benzodiazepines, sold on the street. Despite three decades of effort, there is still no medication therapy for cocaine. Cocaine immunotherapy (popularly called a cocaine "vaccine") to prevent cocaine molecules from entering the brain is now in development, but previews do not look promising for wide-scale use. Other types of medications include blocking agents, such as naltrexone for opiate addiction, which occupy neuronal receptors and blunt a drug's effect. Aversive agents, such as Antabuse (disulfiram), cause people to feel nauseated and vomit when they ingest alcohol. They can be effective in some cases, although many individuals elect to stop taking them.[44]

These medications are not the product of modern neuroscience; they were developed decades ago. Even a vaccine was sought in the 1970s, although today's techniques are vastly more sophisticated. More recently, neuroscientists have collaborated with pharmacologists to develop medications to reverse or compensate for the pathological effects of drugs on the brain. The premise is that different components of addiction can be targeted by different medications. These components are the "reward" circuit (which mediates a strong desire to use and preoccupation with imminent use) and the craving mechanism associated with conditioned cues. Thus far, success has been elusive. Anticraving agents have shown some promise for alcoholics, but treatments for cocaine addiction have been disappointing.[45]

Traditionally, pharmacologists have approached the treatment of alcoholics and addicts in the same way they address most psychiatric diseases: as a matter of reversing or compensating for neuropathology—in this case, the neural alteration resulting from repeated use. This is a logical approach, but instead of focusing almost exclusively on what is wrong in the brain, perhaps they should also investigate the ways in which addicts recover. Addicts find nondrug sources of interests and gratification that generate their own outpourings of dopamine; they practice self-binding and mindfulness exercises that make the prefrontal cortex better at controlling impulses. Relinquishing drugs and alcohol is accompanied by a shift in the brain's valuation systems. How, and even whether, these dynamics will translate into pharmacotherapy is a complicated question, but perhaps the answer will spur discovery of more effective medications—not panaceas but helpful aids to hasten the process of recovery.

Some proponents of the brain-disease model would say that emphasizing the role of choice in addiction is just another way to stigmatize addicts and justify penal responses over therapeutic ones. To this way of thinking, if we see the addict as a "chronic illness sufferer," we will no longer view him or her as a "bad person." This sentiment echoes throughout the addiction community. "We can continue playing the blame game," said Volkow in 2008. "Or we can parlay the transformative power of scientific discovery into a brighter future for addicted individuals."[46]

Sick brain versus flawed character? Biological determinism versus bad choices?[47] Why must these be our only options? This black-and-white framing sets a rhetorical trap that shames us into siding with the brain-disease camp lest we appear cruel or uncaring. The bind, of course, is that it is impossible to understand addiction if one glosses over the reality that addicts do possess the capacity for choice and an understanding of consequences. Forcing a choice between "sick or bad" adds confusion, not clarity, to the long-standing debate over just how much to hold addicts responsible in ways that are beneficial to them and to the rest of society.

Although it makes no sense to incarcerate people for minor drug crimes, exempting addicts from social norms does not ensure them a brighter future. Stigmatization is a normal part of social interaction—a potent force in shaping behavior. Author Susan Cheever, a former alcoholic, coined a new word, "drunkenfreude," to denote how the embarrassing antics of intoxicated friends and strangers keep her sober. "[Watching] other people get drunk helps me remember," Cheever writes. "I learn from seeing what I don't want and avoiding it."[48]

Too often, well-meaning family members and friends try to insulate individuals from the consequences of their behavior and thereby miss an important opportunity to help the addict quit. There is nothing unethical—and everything natural and socially adaptive—about condemning reckless and harmful acts. At the same time, because addicts are people who suffer, we must also provide effective care and support progressive approaches, such as Project HOPE. If we want to garner social and political support for addicts' plight, the best way to do that is to develop the most effective modes of rehabilitation possible—not to advance a reductive and one-dimensional version of addiction.

And what of the efforts to destigmatize addiction through medicalization? Results are mixed. In some surveys of the public, well over half of respondents saw addiction as a "moral weakness" or "character flaw." In others, over half to two-thirds classified it as a "disease." An Indiana University study asked over six hundred people whether they viewed alcoholism as the result of a genetic problem or chemical imbalance (i.e., a "neurobiological conception") or as an outgrowth of "bad character" or "the way he or she was raised." Those endorsing a neurobiological explanation rose from 38 percent in 1996 to 47 percent in 2006; the proportion endorsing psychiatric treatment increased from 61 percent to 79 percent.[49]

Another study revealed an unexpected pattern over the past few decades. As people accepted a biological explanation for mental illness and substance abuse, their desire for social distance from the mentally ill and addicted increased. Biological explanations also appear to foster pessimism about the likelihood of recovery and the effectiveness of treatment.[50] This finding may seem counterintuitive. One might think that a biological explanation would be good news to a patient—and to be sure, some people with mental illness do indeed find it a relief. But when the patient's affliction is addiction and there are no medical cures to restore an addict's disrupted brain, emphasizing the biological dimension seems misguided.

THE authors of the chronic-brain-disease narrative were inspired by discoveries about the effects of drugs on the brain. The promise of finding powerful antiaddiction medications seemed great. The maturing science of addiction biology would mean that once and for all, the condition would be taken seriously as an illness—a condition that began with the explicit, voluntary decision to try drugs but transitioned into an involuntary and uncontrollable state. This knowledge, they hoped, would sensitize policy makers and the public to the needs of addicts, including access to public treatment and better private insurance coverage. A softening of puritanical attitudes and an easing of punitive law enforcement were also on the agenda.

The mission was worthy, but the outcome has been less salutary. The neurocentric perspective encourages unwarranted optimism regarding pharmaceutical cures and oversells the need for professional help. It labels as "chronic" a condition that typically remits in early adulthood. The brain-disease story gives short shrift to the reality that substances serve a purpose in addicts' lives and that neurobiological changes induced by alcohol and drugs can be overridden.

Like many misleading metaphors, the brain-disease model contains some truth. There is indeed a genetic influence on alcoholism and other addictions, and prolonged substance use often alters brain structures and functions that mediate self-governance. Yet the problem with the brain-disease model is its misplaced emphasis on biology as the star feature of addiction and its relegation of psychological and behavioral elements to at best supporting roles. "If the brain is the core of the problem, attending to the brain needs to be a core part of the solution," as Leshner put it.[51] The clinical reality is just the opposite: The most effective interventions aim not at the brain but at the person. It's the *minds* of addicts that contain the stories of how addiction happens, why people continue to use drugs, and, if they decide to stop, how they manage to do so. This deeply personal history can't be understood exclusively by inspecting neural circuitry.

In the end, the most useful definition of addiction is a descriptive one, such as this: Addiction is a behavior marked by repeated use despite destructive consequences and by difficulty quitting notwithstanding the user's resolution to do so.[52] This "definition" isn't theoretical; it explains nothing about why one "gets" addiction—and how could it offer a satisfying causal account when there are multiple levels at which the process can be understood? Our proposed definition merely states an observable fact about the behavior generally recognized as addiction. That's a good thing because a blank explanatory slate (unbiased by biological orientation or any other theoretical model) inspires broad-minded thinking about research, treatment, and policy. Is there room for neuroscience in this tableau? Of course. Brain research is yielding valuable information about the neural mechanisms associated with desire, compulsion, and self-control—discoveries that may one day be better harnessed for clinical use. But the daily work of recovery, whether or not it is abetted by medication, is a human process that is most effectively pursued in the idiom of purposeful action, meaning, choice, and consequence.

This chapter and the preceding one have focused on the biology of desire. We asked whether knowledge of how the brain processes our wants and needs can be applied to the marketplace and in substance-treatment clinics. In both instances, we discovered that although neuroscience research has taught us a good deal about the brain mechanisms underpinning choice making, applying this information to the real world is limited because there are so many levels of influence on human behavior beyond that of the brain. Next we turn to novel lie-detection approaches that interrogate the brain directly. We explore how well brain-based information allows researchers to infer the contents of the mind with regard to deception. We'll discover that discerning truth or dishonesty is not a straightforward matter of brain reading.

4

THE TELLTALE BRAIN

Neuroscience and Deception

IN JUNE 2008, twenty-four-year-old Aditi Sharma was sentenced to life in prison for killing her former fiancé, Udit Bharati. The two had been business students at the Indian Institute of Modern Management in Pune until late 2006, when Sharma dropped out to elope with another man, a fellow student at the management institute. Witnesses claimed that Sharma persuaded Bharati to meet her at a shopping mall, where she offered him a traditional offering known as a prasad, a food item blessed by a Hindu deity. Two days later, Bharati was dead, poisoned with arsenic. Aditi Sharma and her lover were pronounced guilty by a Pune sessions court for conspiring to murder a fellow student.

During her trial for murder, Sharma endured a procedure known as a brain electrical oscillations signature (BEOS) test, claimed to be a neurological lie detector. Similar to EEG technology, the test works by monitoring electrical activity in the brain. Forensic specialists in India, one of the few countries, if not the only one, in which BEOS had been accepted as evidence, claimed that it could determine whether an individual was concealing knowledge of a crime, knowledge that only a guilty person could possess. During an interrogation, investigators would present the accused with facts about the crime, such as the type of weapon used or what the victim was wearing. If the accused person recognized an accurate statement, the electrical detection equipment would register a characteristic brain blip called the P300 wave. The "P" stands for positive, and 300 refers to the fact that the response peaks between 300 and 500 milliseconds after the subject is presented with a stimulus but before he or she is aware of it—and thus that person cannot change the response to it.[1]

In preparation for her BEOS exam, Sharma was fitted with a tight cloth cap studded with thirty-two electrodes wired to a computer. She then sat alone in a room, with her eyes closed, listening to a set of statements prerecorded by the police. Interrogators told Sharma not to respond verbally; her brain would speak for her. A voice on a tape made first-person statements, such as "I bought arsenic," "I met Udit at McDonald's," and "I gave him sweets mixed with arsenic." Sharma insisted that she was innocent, yet her brain reportedly emitted P300 spikes in response to details about the crime. The forensic examiners took this finding as indisputable evidence that she had "experiential knowledge" of the crime and therefore had killed Bharati. The judge handed down a life sentence. It was the first time anywhere that a court had based a conviction on this new lie-detector test.[2]

Sharma's conviction set off a furor outside Indian forensic circles. "The fact that an advanced and sophisticated democratic society such as India would actually convict persons based on an unproven technology is . . . incredible," exclaimed J. Peter Rosenfeld, a psychologist and neuroscientist at Northwestern University and one of the early developers of EEG-based lie detection. In fact, over 160 suspects, in addition to Sharma, underwent testing between 2003—the year BEOS was initially adopted by some Indian police departments—and 2009. It remains legal in Indian courtrooms so long as a defendant consents. In response to Sharma's case, the media warned of "neurocops," "thought police," and "brainjackers." Experts across the globe were outraged by the refusal of Champadi R. Mukundan, the Indian neuroscientist who developed BEOS, to allow independent scientists to review his research protocol and data for potential omissions and errors.[3]

In India, too, some officials were concerned. The Directorate of Forensic Science Services at the Ministry of Home Affairs asked the Indian National Institute of Mental Health and Neurosciences to review BEOS analysis. In a coda to the case, within six months of Sharma's sentencing in 2008, the same forensic lab that "proved" her guilt provided evidence that convicted two other people of the murder of her ex-fiancé. In April 2009 the Bombay High Court relied upon this analysis to conclude that BEOS was "unscientific and should be discontinued," in announcing that Sharma was to be released on bail because of the possibility that evidence had been planted in her purse. Her lover, now her husband, was also released on bail.[4]

Old-fashioned detective work raised the question whether she had even possessed poison. As of late 2012, her appeal was pending. Sharma's fate may remain in limbo for years because of India's slow-moving courts.[5]

The prospect of reading the mind for lies has attracted considerable attention. In the United States, after a decades-long search for an effective lie detector, the effort accelerated in the wake of the September 11, 2001, terrorist attacks. An effective lie detector could revolutionize national intelligence operations, not to mention courtroom proceedings and police work. Thus grants from U.S. agencies, such as the Departments of Defense and Homeland Security, have flowed to university-based investigators. And in the controlled confines of the laboratory with fully cooperative subjects, brain-based lie detection is proving fairly accurate—at least more so than the standard polygraph. To capitalize on what seems like a promising development, two companies have sprung up to market fMRI lie detection to a potentially vast clientele—No Lie MRI in Tarzana, California, and Cephos Corporation, near Boston. "From individuals to corporations to governments, trust is a critical component of our ability to peacefully and meaningfully coexist with other persons, businesses, and governments," says No Lie.[6]

But challenges—and perils—loom large.[7] The first, plainly, is to determine whether one can infer deception in real-world settings from brain scans. The second is to keep immature technology from finding its way into routine use and implicating innocents like Aditi Sharma. And the third is to consider how the courts and society will address concerns about privacy that accompany technological access to our thoughts, feelings, and memories. "Brain privacy" is not under threat—and probably will not be in the near future—so in this chapter we concentrate on the scientific soundness of brain-based lie detection. But we also offer a preview of the constitutional issues surrounding challenges to so-called cognitive liberty.

ONE of the biggest misconceptions about liars is that they give themselves away inadvertently. The ancient Greeks developed a science of physiognomy to identify "tells," involuntary signals, such as a twitch or flush, that supposedly unmask a liar—the same technique that poker players use to determine whether an opponent is bluffing. Historians credit the renowned Greek physician Erasistratus (300–250 BCE) with discerning the concealed love of a stepson for his father's wife by measuring the son's rapid pulse in her presence. Freud thought that anyone could spot deception by paying close enough attention, since the liar, he wrote, "chatters with his finger-tips; betrayal oozes out of him at every pore." People in almost every culture believe that one can spot liars through various cues—the way they avert their eyes, stutter, fidget, or touch their faces. Yet research does not support the validity of these signs; there are surprisingly few useful clues for detecting when people are lying. And even these cues are primarily verbal rather than nonverbal; for example, inaccurate statements are somewhat more likely to contain fewer details and more qualifiers ("I'm not positive, but I think that the bank-robber's shirt may have been blue") than accurate ones. Even trained security professionals, such as judges and police officers, rarely do better than chance at detecting lies.[8]

Our incompetence at detecting lies is a liability in a world so full of them. People admit to lying in about one in every five social interactions lasting more than ten minutes—at least once a day, on average. According to one dogged soul who searched the literature, the English vocabulary contains 112 differently shaded words for deception: "collusion," "fakery," "malingering," "confabulation," "prevarication," "exaggeration," "denial," and so on. The late British psychiatrist and deception expert Sean Spence observed that across cultures, there are more words for deception than for honesty, perhaps because there are many ways to deceive but only one way to tell the truth.[9]

This is no surprise, really. Deceiving one another is an essential part of social life. We cooperate with others by artfully manipulating relationships and misleading our competitors. Coupling sometimes relies on these strategies, as anyone who has been the target of a talented seducer (or has been a talented seducer) can attest. Our deceptions are made possible by our ability to see the world through others' eyes and anticipate their actions. Philosophers and psychologists call this capacity "theory of mind." Most children begin to develop this understanding between the ages of three and four; the better children are at intuiting the desires, intentions, beliefs, feelings, and knowledge of others, the more effective they are at deceiving parents, teachers, and friends.[10]

For about a century, the famously flawed polygraph test has been the staple technology of lie detection. Its technology reflects the long-held assumption that lying is stressful for the deceiver and that such stress will manifest itself as high blood pressure, rapid breathing, or sweaty palms—reactions produced by the peripheral nervous system. A primitive application of the theory can be seen in ancient China, where interrogators made people accused of crimes hold rice in their mouths or swallow dry bread. If the rice remained dry or the bread did not go down easily, the suspects were deemed guilty. In this line of thinking, deception led to anxiety—fear of getting caught, distress over having betrayed someone, and guilt over having violated one's moral standards—and that caused dry mouth.[11]

In the early 1900s, William Moulton Marston, an undergraduate at Harvard, invented the precursor of the modern polygraph. The device recorded breathing rate by means of a pneumatic rubber hose wrapped around the subject's chest and a blood-pressure cuff encircling the upper arm. In a charming footnote to polygraph history, Marston later became a comic-book writer and, under the pen name of Charles Moulton, created Wonder Woman, an action heroine who wore a "Golden Lasso of Truth" around her waist. When villains were lassoed with her magical version of the pneumatic hose, they were forced to tell the truth.[12]

The polygraph has led a less enchanted life. The official beginning of its history of legal and scientific controversy can be traced to 1923, when a federal court ruled that results from the test were inadmissible as evidence because Marston's technique had not gained general acceptance within the scientific community. This ruling, in *United States v. Frye*, was a landmark in evidence law because it provided the first clear judicial statement of the standards for scientific evidence. Under the *Frye* standard, as well as the more recent *Daubert* standard, which has replaced *Frye* in federal courts and most states, polygraph evidence has been excluded from trial in nearly every state and federal court for the past ninety years. Outside the courtroom, however, the polygraph became a routine feature of American law enforcement. By midcentury, the device was being used to safeguard nuclear secrets, assure the political fidelity of scientists, and purge homosexuals from government jobs.[13]

The 1988 Employee Polygraph Protection Act bans private employers from using polygraphs in preemployment screening or in trying to ferret out theft by workers. A decade after the law's enactment, the U.S. Supreme Court ruled that state and federal governments may ban the use of polygraph evidence even if a defendant insists that the results would vindicate him or her. Courts remain leery of the policy. Some federal circuits and a handful of states continue to admit polygraph evidence under special circumstances, but only the New Mexico courts make it generally admissible. Outside courtrooms, however, national security and law-enforcement agencies perform more than 1 million polygraphs each year in the United States just to screen potential employees or clear workers before advancing them to more sensitive positions.[14]

To see why the polygraph has come in for such scrutiny, one needs to know how it works. Let's say that our suspect stole $5,000. Under the standard polygraph procedure, an interrogator asks three types of questions. To set a physiological baseline for honest responses, he instructs the suspect to reply honestly to "irrelevant" questions, such as "Do you speak English?" or "Is it October?" He also asks "control" questions about past minor infractions—"Did you ever get a traffic ticket?" "Receive too much change from a cashier and keep it?" "Tell your boss a lie?" Almost all of us have received a parking ticket, pocketed an extra buck, or told a workplace fib at least once, but because we wouldn't want to admit these peccadilloes during a polygraph test, we'd presumably need to lie and thus trigger a mild perturbation of heart rate or sweatiness of the palm. The polygrapher uses these "control" questions to set a "white-lie" baseline against which more relevant, crime-related lies—and their presumably greater states of physiological arousal—are compared. Thus, when guilty suspects answer "no" to the interrogator's "probe" question "Did you steal the money?" they will display a more robust physiological response than when they bend the truth in a white lie. Conversely, if they are innocent, a "no" response to the question should elicit a weaker physiological response relative to that for the white lie.

There is a comforting logic here—if you are guilty, your body will give you away—but it is at best wildly oversimplified, at worst it is patently false. Habitual liars are not necessarily anxious; this is especially true of psychopaths, whose peripheral nervous systems are less responsive to threat than are most individuals'. And truth tellers, for their part, sometimes are anxious, especially when being questioned in a high-stakes situation.[15] To a lie-detector machine, innocent people often *seem* guilty. Under interrogation, they become frightened or agitated, their hearts pound, their breath labors, and their palms sweat. They may even *feel* guilty. Polygraphers dub these people "guilt grabbers" because the mere thought of being accused of wrongdoing gooses their autonomic nervous systems. Conversely, guilty people, who are often practiced criminals, often know how to beat the polygraph. They might bite their tongue hard or engage in strenuous mental arithmetic while answering the white-lie questions to set off physiological reactions. Thus, when they lie about the actual crime, their results are less dramatic.

At bottom, then, the polygraph is an arousal detector, not a lie detector. It is prone to generating high rates of "false positives," which can lead authorities to punish the innocent, and to a somewhat lesser extent, "false negatives," which wrongly exonerate the guilty. The National Academy of Sciences estimates that well-conducted polygraph exams correctly identify roughly 75 percent to 80 percent of those who lie (true positives) but also mislabel as liars about 65 percent of truth tellers (false positives). Two of the most famous errors are the failure in 1986 to implicate the guilty Aldrich Ames, a CIA agent who spied for the Soviets (a false-negative error), and the misidentification in 1998 of Wen Ho Lee, a Department of Energy scientist, as an agent of the Chinese government (a false-positive error).[16]

IF the body can't be trusted to reliably betray its secrets, would going straight to the brain, the organ of deceit, be a better way to reveal deception? There are two basic approaches, both of which rely on EEG or fMRI to detect deception. One way is to see whether suspects are keeping information to themselves. The guilty knowledge test (GKT) targets such sins of omission.[17] The other strategy is to identify brain activity that enables us to distinguish lying from truth telling. Like the polygraph, brain-based lie detection asks the basic inquisitorial question, "Did you do it?" The GKT simply requires suspects to have a memory of the crime and in essence asks, "Do you recognize these facts of the crime?"

More specifically, the GKT presents details to a suspect that would be known only by someone who is guilty. Thus the interrogator might ask, "What was the caliber of the gun you used? Was it .22, .25, .38, or .44?" or "Where was the family safe located? Behind a bathroom mirror? In the basement? Behind a bookcase?" Suspects who consistently exhibit stronger physiological responses to the correct option (e.g., the actual caliber of the gun, the safe's true location) probably possess incriminating knowledge. Conversely, people who react to all the options with equal intensity are probably innocent. A control reading is obtained by presenting the suspect with information about the crime that anyone would know from reading news accounts. An "irrelevant" prompt is also presented, such as asking the suspect which date is meaningful after having inserted his or her birthday among a group of random dates. The virtue of the GKT under controlled conditions is that the false-positive rate is low, and, under well-defined laboratory conditions, the test is quite accurate. The problem, according to many critics, is that its current most outspoken promoter, psychologist Lawrence A. Farwell, has taken great liberties with it.

In 2001, just weeks after the al-Qaeda terrorist attacks, *Time* magazine revived interest in the GKT by putting Farwell in its Top 100 "Innovators Who May Be the Picassos or the Einsteins of the 21st Century." He had developed a technology he called "brain fingerprinting," which used brain waves to identify guilty knowledge. According to *Time*, "Farwell believes that he can determine if a subject is familiar with anything from a phone number to an al-Qaeda code word." For several years, Farwell had been in contact with federal agencies such as the Central Intelligence Agency and the Secret Service about using brain fingerprinting for military and security purposes. Farwell's Brain Fingerprinting Laboratories is located in Seattle. Brain fingerprinting uses a controversial electrical marker of recognition that Farwell calls MERMER (Memory and Encoding Related Multifaceted Electroencephalographic Response), of which the P300 wave is a major component. If this sounds like BEOS, it is no coincidence; Farwell's work served as the inspiration for the test used to interrogate Sharma.[18]

Forensic psychologists have charged Farwell with serious over-claiming. He has published little in peer-reviewed journals and has refused to make his work available for independent review. Very much the showman, Farwell and a film crew from ABC's *Good Morning America* traveled to Oklahoma in 2004 to test, on camera, a death-row inmate named Jimmy Ray Slaughter. Farwell claimed that Slaughter's brain manifested no spikes of recognition when the convict was presented with the correct answers, suggesting that he was innocent by brain fingerprinting standards. But the appeals court justices refused to grant him an evidentiary hearing, and Slaughter was executed in 2005.[19]

Although Farwell and the Indian investigators who interrogated Aditi Sharma with BEOS used EEG as the basic brain-assessment technology, some investigators have tested the guilty knowledge paradigm using fMRI. Instead of examining brain waves, the investigators present subjects with elements of the crime scene and then look for a pattern of blood oxygenation level dependent (BOLD) signals in brain regions related to memory that suggests previous experience with the scene. No matter which technique is used in the guilty knowledge approach, the neural representation of memory (as a crude brain-wave blip or a more nuanced pattern of brain activation) is the essence of the guilty knowledge approach—and also its Achilles' heel.

Farwell claims that brain fingerprinting detects whether particular information is "stored in the brain." But this is a flawed metaphor for how memory works. The brain does not act like a faithful audio-video recorder, nor is it a repository for static recollections. Memory is a fallible instrument, sometimes spectacularly so. Not everything is remembered, and what is remembered is often distorted. At each stage of memory—encoding the event, storing it, creating a permanent record, or retrieving it—something can go awry. People who commit crimes might nonetheless "pass" a brain-wave interrogation because, in the heat of passion or rage, they did not note crucial details of the crime. And if something goes unnoticed, the brain cannot encode a memory. Even when details are encoded, they are not always stored permanently. They can undergo normal decay, or, over time, become contaminated by both earlier and later memories. Such composite memories can seem as vivid and powerfully real as accurate ones.[20]

Counterfeit memories are difficult to distinguish from memories of real events. This is a well-known bane of eyewitness identification and forensic interviewing, especially with suggestible children. Psychologists at the University of Arizona induced false memories in subjects to determine whether they would appear the same as true ones under a P300 paradigm. Using a well-established psychological test, they read to subjects a series of related words—"prick," "thimble," "haystack," "thorn," "hurt," "injection," and "syringe." The word "needle"—a natural fit with the others—was not included. Yet when the investigators asked subjects whether it had been one of the words they had heard moments earlier, many participants answered yes. In P300 testing, those who reported feeling confident that "needle" had been among the original words showed the same pattern of brain electrical activity as they did when they were recalling the words they had heard. It turns out, in short, that the GKT is more a belief-meter than a truth-meter.[21]

The same phenomenon has been demonstrated using fMRI, confirming earlier findings that imagery and perception share common processing mechanisms. Psychologist Jesse Rissman and his colleagues scanned the brains of subjects as they memorized over two hundred faces and processed the data using pattern recognition, or "decoding," software. Under their technique, a subject observed images in succession while his brain activity was transferred to a high-speed computer, which "learned" what each memorized face looked like in terms of a unique "neural signature." An hour later, researchers showed subjects the same faces interspersed with previously unseen faces, for a total of four hundred images. The findings were striking: The researchers could not differentiate neural signatures associated with the faces seen by the subjects from those elicited by faces the subjects were seeing for the first time yet thought were familiar. This important study underscores a major limitation of fMRI in distinguishing a true from a false memory, a formidable hurdle to using brain-based evidence in judicial settings.[22]

Unreliable memories can lead to false negatives on the GKT, but false positives can happen too, because the P300 response (or an fMRI neural signature) is not specific to guilty knowledge. Just as an innocent person undergoing a polygraph may have sweaty palms or an elevated pulse because he or she feels anxious, the most that can be said of a P300 response is that it reflects the recognition of something special and recognizable to the subject. In the case of a visual cue, such as the gun used in the crime, a P300 surge to a given stimulus could reflect the fact that a suspect has read about the weapon and vividly imagined it or has seen such a gun before in another context.

Finally, the GKT faces imposing logistical hurdles. The crime scene must remain largely or entirely untouched before investigators arrive. If the environment is disturbed and the information used to construct multiple-choice probes is incorrect, guilty suspects will not display recognition-related arousal and may appear innocent. Conversely, if details are leaked to the media, an innocent person who follows the news might show signs of recognition and appear guilty. In addition, investigators must have access to enough separate pieces of specific information about the crime scene and the nature of the crime to construct a meaningful multiple-choice test. For all these reasons, the GKT remains a clever investigative tool, but one that works best within the controlled confines of the laboratory.

Now we turn to the second and more popular kind of brain-based approach to lie detection: the neural lie detector, based on the idea that different brain systems are invoked during lying as opposed to truth telling. If researchers could pinpoint the specific neural correlates of a lie using fMRI, this discovery could be the holy grail of deception detection. The prevailing theory of fMRI-based lie detection is that specific brain regions work harder when people lie—presumably because the brain must first inhibit honesty and then generate a falsehood.[23] Subtracting fMRI signals associated with lying from those associated with truth telling should, in theory, reveal the neural signature of deceit. Put differently, what fMRI detects, according to this model, is the neural representation of conflict between a dishonest and a truthful condition.

In 2005, psychiatrist F. Andrew Kozel published one of the most cited experiments in fMRI-based lie detection. He and his colleagues recruited volunteers to participate in a so-called mock-theft paradigm. In the study, researchers escorted subjects, one by one, into a room containing a desk and instructed them to take an item from the desk drawer—either a ring or a watch—and put it in a nearby locker. Before being scanned, they received important instructions to deny having "stolen" the watch or ring when asked. This meant that subjects would always press the "no" button in response to the questions researchers flashed on the scanner's computer screen: "Did you take the watch?" and "Did you take the ring?"[24]

In this clever manner, subjects' responses would inevitably be truthful in one instance and untruthful in the other.[25] Researchers then subtracted both the truth and lie conditions from a neutral baseline of activity that they had earlier established. The results for all subjects were pooled to generate a composite map showing seven brain regions that were more active when subjects lied than when they told the truth. These results, however, said nothing about individual subjects—so how could the researchers tell when a specific participant was lying?

In the second part of the experiment, the team recruited a second group of volunteers to participate in the same mock-theft experiment. The researchers then compared the imaging results of this second set one by one against the composite image they obtained earlier from the first set. This allowed them to determine whether a given subject took the ring or the watch with 90 percent accuracy. (Other mock-theft experiments produced less eye-catching results than Kozel's, with accuracy rates hovering between 70 percent and 85 percent.)[26]

This general-subtraction method is the basis of the recent legal use of fMRI lie detection. In 2009, a custodial father on trial for child abuse hired No Lie MRI to prove that he did not have sex with his daughter. According to the No Lie MRI report submitted to the San Diego Juvenile County Court, the man's "no" responses to such questions as "Did you have oral sex with X?" were truthful. Ultimately, the defense withdrew its request to introduce the fMRI evidence after the prosecution retained an expert who would have testified against fMRI-based lie detection.[27]

In a higher-profile case the following year, fMRI lie detection came under intense scrutiny. The federal government charged Tennessee psychologist Lorne Semrau with defrauding Medicare and Medicaid of several million dollars between 1999 and 2005. Semrau said that he was confused by the claim-filing process but maintained that he never intended to steal. His lawyer hired the other fMRI lie-detection service, Cephos, to investigate his past mental state. The test questions included "Did you bill [medical billing code 99312] to cheat or defraud Medicare?" Cephos concluded that Dr. Semrau's "brain indicates that he was telling the truth" when he said that he did not intend to cheat. Before trial, the prosecution objected to the admission of this evidence, so the judge held a pretrial hearing to assess the scientific validity of fMRI-based lie detection. At this so-called *Daubert* hearing, experts testified for and against the soundness of Cephos's data.[28]

In the end, the judge ruled that the defense could not present fMRI evidence in court because the error rate (the probability of wrongly detecting a lie in someone who was telling the truth or of missing a lie in a liar) was unknown and because the scientific community had not yet accepted it as a valid technique.[29] A federal appellate court upheld this ruling in the fall of 2012. Judges also refused to admit fMRI-based lie-detection evidence in two other cases: an employment discrimination lawsuit in New York City in 2010 and a rehearing in a murder case in Maryland in 2012.[30]

In all these instances, the soundness of science was at issue. As impressive as some of the lab studies were, judges found little justification for believing that the technique was as accurate outside the lab. And with good reason, because many factors can affect the neural correlates of lying.

First, consider the difference between "lab lies," the kind that Kozel and other researchers elicited and tried to identify, and real lies. Most obvious, no one tells real suspects in forensic settings to lie, let alone to lie in particular ways. Intent to deceive is so integral to the phenomenon we call a lie that many neuroscientists contend that the subjects in experiments are committing not a lie but an "instructed falsehood." Instructed falsehoods and purposeful attempts to mislead almost certainly make different demands on the brain, raising the further question of what exactly fMRI is measuring in these studies. Finally, most lab subjects are happy to go along with the testing, whereas real suspects might try to beat the machine by moving their heads, humming, or silently performing multiplication in the hopes of distorting the imaging signal. In one study, investigators found that simply wiggling a single finger or toe could reduce the accuracy of lie detection from near perfect to one-third.[31]

Second, the neural signatures of real lies almost surely represent more than the lie per se. As neuroscientist Elizabeth A. Phelps points out, an actual suspect accused of a crime faces a highly emotional situation with high stakes.[32] He or she also has time to ruminate and to imagine (if innocent) or recast (if guilty) the event. A guilty suspect also can rehearse a story. This means that the neural signature of a real lie is more than just the representation of conflict between a dishonest and a truthful condition; it also incorporates the neural correlates of emotion and imagery that would not be found in a less fraught lab lie.

Third, consider who is doing the lying in these lab experiments. Participating students, for the most part, don't have mental health problems or old head injuries, and they haven't habitually used drugs. Nor have most even committed or been accused of a serious crime, so any generalization of study findings to the wider population must be made with caution. Nor do they have nearly as much at stake if their lie is caught as would a guilty suspect. But the legal system deals with actual suspects who often have low IQs, histories of substance abuse, previous head injuries, and long criminal records. Their emotional investment in appearing honest is presumably a lot higher. This point is relevant because, as noted, emotion is known to affect neural activation patterns associated with cognitive tasks.

In addition, people who volunteer for studies may not be especially good liars, whereas real-world troublemakers may well be experienced prevaricators whose brains could show less activation when they are lying thanks to extensive practice. Guilty parties accused of a real crime also have time to manufacture a version of events and commit it to memory. Rehearsal, then, is another major difference between lab lies and real lies. Also, guilty people who come to believe their own claims of innocence or have rehearsed alibis could escape detection. In contrast, merely thinking about lying might get an innocent person in trouble. In one study, researchers found that the neural activity associated with thinking about lying regarding the results of a coin toss was indistinguishable from activity associated with actual lying.[33]

Finally, fMRI-based lie detection is undermined by inconsistent results. To be sure, when groups of subjects are compared, the brains of subjects who deceive tend to show different activation and deactivation patterns from those of the brains of truth tellers. Over two dozen studies confirm this conclusion. Yet in no study has a specific brain region or set of regions been identified that is consistently activated in all people when they fib, and consistently silent when they do not. Indeed, the array of brain regions correlated with deception is dizzying: the parahippocampal gyrus, the anterior cingulate, the left posterior cingulate, the temporal and subcortical caudates, the right precuneous, the left cerebellum, the anterior insula, the putamen, the thalamus, and the prefrontal regions (anterior, ventromedial, and dorsolateral), as well as regions of the temporal cortex. Clearly, the enormous variation means that no single neural pattern of activation can presently distinguish deception from truth telling. This makes it difficult, if not impossible, to formulate a reliable "lying brain" signature.[34]

In the aggregate, these caveats—most decisively, perhaps, the failure of studies to re-create the conditions characteristic of real-world lies—should disqualify today's brain-based lie detection from courtroom use.

Now let us add even more nuance by considering the very nature of lies. Scientists who have examined lies per se have found that different types activate different parts of the brain; not all lies are psychologically similar. In their seminal studies, psychologists Stephen Kosslyn and Giorgio Ganis focused on two types of lies: spontaneous lies and rehearsed, or memorized, lies. The latter, as the name implies, are those you are prepared to tell when your friend asks you if you are sticking to your diet. A prepared answer might be "I had a tiny salad" when the truth is that you had a burger and fries. Spontaneous lies are those you tell on the fly, as when your friend asks you whether you can give her annoying boyfriend a ride to the airport, and you say you can't do it because your car is in the shop.

Kosslyn and Ganis hypothesized that when people tell rehearsed lies, they merely need to retrieve them from memory. A spontaneous lie, by contrast, takes more work. When your friend asks you to chauffeur her boyfriend, you must engage episodic memory (responsible for recalling events), such as your past dealings with the boyfriend, and semantic memory (responsible for recalling knowledge) to help manufacture the lie. Presumably, a spontaneous lie would be richer in detail, too, involving visual images or feelings that are encoded in various parts of the brain and thereby giving rise to a more complicated neural representation.[35]

In their experiment, the researchers asked subjects to describe two experiences: their best job and their most memorable vacation. They asked the subjects to choose one of the two experiences, the job or the vacation, whichever they preferred, and to create an alternative version of it and memorize it. So, if the actual vacation was "My parents and I flew from Boston to Barcelona on Continental Airlines and stayed at the Granvia Hotel," the altered version might be "My sister and I drove from Los Angeles to Mexico City and stayed in a hostel." A student memorized the false version for about a week and returned to the lab to be scanned. During the scan, researchers told each student to make up some new (spontaneous) untruths on the fly. So subjects would lie on the spot when asked where they went on vacation and replace Mexico City with, say, Miami or respond "my aunt" when asked who their travel companion was. A parallel scenario was tested for the best job one ever had if students chose that option.

As the researchers predicted, different brain networks were engaged during spontaneous lying than during rehearsed lying, and both differ from those used during truth telling. Both involve memory processing, but when subjects lied spontaneously, their brains drew more heavily on the anterior cingulate cortex, which presumably facilitated the suppression of what otherwise would have been a truthful response. When their lies were rehearsed, a region in the right anterior prefrontal cortex (involved in retrieving episodic memory) was selectively activated. Truthful memories were the least effortful to produce, presumably because they were acquired naturally and did not require the kind of auditing and editing that spontaneous lies required.[36]

The point is this: No brain region uniquely changes activity when a person lies; each type of lie requires its own set of neural processes. This is because lies are not all alike psychologically. Journalist Margaret Talbot offers a nuanced litany of lies based on motive: "small, polite lies; big, brazen, self-aggrandizing lies; lies to protect or enchant our children; lies that we don't really acknowledge to ourselves as lies; complicated alibis that we spend days rehearsing." Some lies are even told for the mere fun of fooling others, a practice psychologists call "duping delight." And what about the "more-or-less honest omissions, exaggerations, shadings, fudgings, slantings, bendings, and hedgings" that are an omnipresent feature of litigation, one scholar asks?[37]

Montaigne, the sixteenth-century French Renaissance essayist, reflected on the kaleidoscopic variety of deception: "The reverse side of the truth has a hundred thousand shapes and no defined limits." Half a millennium later, researchers are beginning to discern some of those shapes. The lies you tell about yourself, for example, look different on brain scans from those you tell about others. A lie about, say, one's house will rely on quite different cognitive functions than a lie about a future home, which will engage its own patterns of thought, emotion, and imagination. A lie that generates profound remorse won't overlap fully, if at all, with the neural correlates of a glib fib. A lie about the future will differ in its neural correlates from one about the past. Montaigne was right: From the whitest of lies to the darkest of deceptions, "the reverse side of the truth has no defined limits."[38]

BRAIN-BASED deception detection can perform impressively in the lab, but there is no evidence that its capabilities extend safely to forensic settings. Nonetheless, No Lie MRI and Cephos have vigorously promoted its use. No Lie entered the detection business in 2006, and Cephos ("Our Business Is the Truth") followed in 2008. No Lie and Cephos foresee a day in the near future when fMRI-based "truth verification," a term both use, will supplant such regular workplace checks as drug screenings, résumé validation, and security background clearances. Most clients, says No Lie president Joel Huizenga, are suspicious spouses who want to prove their partners' fidelity. He does not shy away from making extravagant claims for his fMRI technique: "It doesn't matter whether you feel guilty or not, it doesn't matter if you've memorized your story, and it doesn't matter whether you believe your lie would save the world. We can still spot [the lie]." "The *last realm of privacy* is your mind," says the head of Veritas Scientific, an arm of No Lie. "This will invade that," he states, in reference to a still-under-construction BEOS-like helmet designed to aid military intelligence.[39]

No Lie claims at least 90 percent accuracy for its methodology; Cephos, 97 percent. "What we are able to do," says Huizenga, "is look inside people's brains and verify that they are telling the truth." This is why people like Harvey Nathan pay $5,000 to $10,000 to undergo MRI lie detection. Nathan, of Charleston, South Carolina, hired No Lie MRI in 2007 to prove to his insurance company that he had not burned down his deli four years earlier. Although Nathan was cleared of arson in a criminal case, his insurer remained unconvinced and withheld payment. After several years of haggling, Nathan flew to Los Angeles to be scanned by No Lie MRI. According to the test, Nathan was telling the truth about not setting the fire, but as of late 2011, he reported that he had not received a settlement from the insurance company.[40]

Although commercial companies have not yet succeeded in getting their evidence into trial, they remain sanguine. "Cases will come to court, they just have to come to the right venue," said Huizenga of No Lie after its report was not introduced in the San Diego case of alleged child abuse. When the judge ruled against the admissibility of the Cephos report in the Tennessee fraud trial of Dr. Semrau, Cephos president Steven Laken, too, was undeterred. "This is just one ruling," he said.[41]

Right now it seems that the best current brain-based modes of deception detection can do is play off the public's belief that they might work. In the case of the iconic polygraph, people tend to have so much faith in its imagined infallibility that examiners sometimes play up the ritual of the test to trick people into disclosing information. President Richard M. Nixon understood the advantage of this fear when he considered submitting hundreds of government employees to polygraph tests to pinpoint the source of news leaks about international treaty negotiations. "I don't know anything about polygraphs," he told an aide, "but I know they'll scare the hell out of people." Nixon's rationale has been confirmed by ample research demonstrating that when people are hooked up to a fake but realistic-looking apparatus (wonderfully dubbed the "bogus pipeline to truth"), they are likely to tell the truth.[42] Devices based on fMRI, given the impressive technology required, could be even more effective in duping the public about their effectiveness.

The late psychologist David P. McCabe and colleagues designed an experiment to test whether subjects found fMRI evidence more influential in determining guilt than evidence from other lie-detection techniques. They asked subjects to determine the guilt of a man accused of killing his estranged wife and her lover. To instill reasonable doubt, researchers told them that evidence against the man was "incomplete and ambiguous." In addition to an fMRI scan of the defendant, researchers provided polygraph data on him and evidence from a controversial new technology called facial thermal imaging. Briefly, thermal imaging (TI) measures heat levels across a person's face and represents the information visually in the form of a colorful scan. The premise is that facial blood vessels dilate and thereby release heat when a person lies. In the end, McCabe found that subjects who deemed the man guilty and relied heavily on scientific evidence to make that determination accorded fMRI significantly more weight than they did the polygraph or the TI. McCabe concluded that fMRI's persuasive power came not from its novelty or its visual product—after all, TI evidence was both novel and visual—but rather because it purported to provide information that came directly from the brain.[43]

Should deception detection eventually overcome the myriad technical obstacles in its way, it would still face close scrutiny. Civil libertarians are wary of potential violations of mental privacy and "cognitive liberty." "We view techniques for peering inside the human mind . . . as a fundamental affront to human dignity," says an American Civil Liberties Union spokesperson. Although "mental privacy panic," as one legal scholar called it, is unwarranted, safeguards have been proposed. Some ethicists and neuroscientists have called for regulation and preapproval of deception-detection technology in much the same way as the Food and Drug Administration requires two sets of randomized, controlled trials to approve new drugs. Others have called for the creation of a national advisory committee on neurosecurity to advise cabinet departments on minimizing the misuse of biological research.[44]

The possibility of effective lie detection interests constitutional law scholars as well. Of particular interest are the implications for the Fifth and Fourth Amendments. Let's first consider the Fifth Amendment, which protects a suspect's right to remain silent lest he inadvertently bear witness against himself. The U.S. Supreme Court recognizes two classes of potentially self-incriminating information: physical and testimonial. Physical evidence, such as blood, hair clippings, and DNA samples, may be compelled during a criminal investigation; testimonial evidence, such as statements and other communicative acts such as nodding, cannot.[45] Perhaps one day, the court will face the question whether neuroimaging is physical and unprivileged or testimonial and privileged.

There's no clear answer because neither "physical" nor "testimonial" accurately describes brain-derived evidence, says legal scholar Nita Farahany. Brain-derived information is both testimonial, because it reveals mental contents (albeit imperfectly), and physical, because it represents a person's thoughts in terms of blood oxygen levels or brain waves. Paradoxically, the suspect could remain silent, yet the state could potentially extract information directly from the brain in ways he or she could not control.[46]

The courts will also be confronted with challenges stemming from the Fourth Amendment. This amendment safeguards the right of people to be secure against unreasonable searches and seizures by the government. The relevant question here is whether obtaining brain-based evidence is tantamount to a search under the Fourth Amendment—that is, does it elicit information that would be unknown but for the search itself—or is it more like capturing ordinary physical evidence, such as saliva on a cigarette butt?[47] The interior of one's skull would certainly seem a place where one would have the right to expect privacy.

As we've seen, there are formidable challenges to drawing accurate inferences about the deceiving mind from information derived from the brain. Under controlled conditions, the guilty-knowledge-test and mock-theft paradigms have yielded impressive results. Yet the limits to identifying real-world deception with its emotional components remain great—so great that premature application risks harming the innocent and perhaps exculpating the guilty. It also risks

misleading the clients of companies such as No Lie MRI and Cephos, who have placed faith in their capacity to accurately appraise
their honesty or the honesty of others.

Just as there is no brain region or circuit that uniquely reflects
lying, there is almost certainly no single signature of a guilty brain.
Defense lawyers, especially in death-penalty cases, are increasingly
turning to brain imaging to provide evidence bearing on their clients'
rational capacities, intentions, and ability to tell right from wrong. By
doing so, they hope to mitigate their clients' punishment or help
them evade criminal responsibility altogether. In the next chapter, we
turn to the fascinating but vexing question of what brain-based technologies can—and cannot—tell us about the minds of people whose
fates hinge on the accurate reading of their brains.

5

MY AMYGDALA
MADE ME DO IT

The Trials of Neurolaw

O N THE AFTERNOON OF SEPTEMBER 9, 1993, two fishermen discovered the body of Shirley Ann Crook floating in Missouri's Meramec River. She had been hogtied with electrical wire, and her face was covered in a towel secured by layers of duct tape. The next day, police arrested Christopher Simmons on suspicion of her murder. The seventeen-year-old high-school student quickly confessed that he and a fifteen-year-old friend had broken into Crook's home soon after midnight two nights earlier. He said that upon entering her bedroom, he was surprised to recognize Crook, forty-six, from a minor car accident in town that had involved them both.[1]

The teens bound and gagged Crook before loading her into the back of her own minivan and driving to Castlewood State Park. Deep in the woods, Simmons and his friend parked near a train trestle that spanned the Meramec. They prodded the whimpering woman up the stairs and retied her when they reached the top. Before dawn, they shoved Crook into the black water forty feet below.

At school the next day, Simmons bragged to his friends that he went through with the murder because "the bitch seen my face." But the decision to kill Shirley Ann Crook was made before he ever set foot in her house. As witnesses told police, Simmons had often spoken to friends of robbing a person, binding him, and throwing him off a bridge. What's more, he had assured his friends that they could "get away with it" because they were minors. But Simmons was badly mistaken. At the time, Missouri was among a handful of states in which it was legal to execute juvenile criminals. In June 1994, as his classmates were graduating from high school, Simmons sat on death row in Missouri's Potosi Correctional Center facing lethal injection.

Eight years later, Simmons's lawyers learned about a case before the U.S. Supreme Court that inspired them to appeal to the Missouri Supreme Court. In that case, *Atkins v. Virginia*, the Court was asked to decide whether executing mentally retarded criminals violated the Eighth Amendment's protection against cruel and unusual punishment. (Daryl Atkins, a twenty-four-year-old man with an IQ of 59—11 points below the standard IQ cutoff of 70 for mental retardation—was on Virginia's death row for killing a man during a robbery.) Atkins's lawyers argued that the "death penalty [is] unacceptable" for people with mental retardation because of their impaired "ability to control their behavior, their understanding of the context in which they behave, the maturity and responsibility with which they reach moral judgments."[2]

Even before the ruling for Atkins came down in May 2002, Simmons's lawyers rushed to appeal to the Missouri Supreme Court to address the constitutionality of executing those who committed a crime before age eighteen. "A good deal of recent research shows strong scientific support for the fact that, biologically, juveniles lack the mental capacity to act with the same level of moral culpability as adults," argued Simmons's counsel. Simmons prevailed. The Missouri court vacated his death sentence and banned juvenile executions more broadly. Simmons, then twenty-seven, was resentenced to life in prison without the possibility of parole.[3]

But the legal saga of Christopher Simmons did not end there. The state moved to proceed with the execution and appealed to the

U.S. Supreme Court to reverse the lower court's decision banning capital punishment for those who committed crimes as minors. In this case, known as *Roper v. Simmons*, the young man's lawyers emphasized teens' diminished capacities, relying substantively on the biological immaturity of the teen brain. "Brain Science v. the Death Penalty," read a headline in the *Boston Globe* in the fall of 2004, the day before the U.S. Supreme Court was to hear Simmons's case. The lawyers' argument turned on relatively new data establishing that human brain maturation continued into the mid-twenties, contrary to earlier understanding that pegged late childhood, about age eleven, as the completion of brain development.[4]

"To a degree never before understood, scientists can now demonstrate that adolescents are immature not only to the observer's naked eye, but in the very fibers of their brains," said a joint amicus brief submitted by the American Medical Association, the American Psychiatric Association, and other groups.[5] The "very fibers of their brains" was no metaphor. As the amici described, brain development entails orchestration among regions that communicate with one another through tracts of neuronal axons, also called fibers. Fibrous pathways run from the frontal lobes, which are associated with impulse control and risk assessment, to the amygdala, which is linked to the primitive impulses of aggression, anger, and fear, among other emotions.

Optimally, the frontal lobes modulate the amygdala, a working relationship that depends on a well-functioning connection between the two. But in teens, the connection is incomplete because the fibers are not yet fully wrapped in myelin, the fatty conducting tissue that speeds the transmission of electrical impulses along axons. Until myelination is finished, the frontal lobes cannot exert as much of a check on the amygdala-mediated emotions as they do in adults.[6]

Teens' frontal lobes are also under construction. Superfluous synaptic connections are being pruned in much the way a gardener cuts back tangled branches—a process believed to allow the remaining neurons to function more efficiently. The teen amygdala is still another work in progress. Its sensitivity to stress and threats makes it a twitchy accelerator in conjunction with the frontal lobes' imperfect brakes. Finally, some researchers believe that the adolescent reward system is more reactive than adults', presumably stoking teenagers' attraction to pleasurable, sensation-boosting activities—including the approval of peers. The amicus brief laid out these changes in detail and warned the court that executing youthful offenders would be tantamount "to hold[ing] them accountable . . . for the immaturity of their neural anatomy and psychological development."[7]

In March 2005, the Supreme Court decided *Simmons*, voting five to four to ban the execution of minors. Some juvenile advocates hailed the case as a modern classic—the *Brown v. Board of Education* of "neurolaw," as one legal scholar put it.[8]

BARELY a decade old, neurolaw, a discipline that sits at the intersection of brain science, legal theory, and moral philosophy, is a rising star on the legal horizon. "Neuroscience could have an impact on the legal system that is as dramatic as DNA testing," said the president of the John D. and Catherine T. MacArthur Foundation, which initiated a $10 million Law and Neuroscience Project in 2007 to explore the implications of brain science for criminal law. Bioethics councils under both Presidents George W. Bush and Barack Obama addressed cognitive neuroscience and its potential to shed light on mental attributes—reason, judgment, and impulse control—that are relevant to legal culpability. The Royal Society in the United Kingdom took up these issues in 2011, and the academic literature on the topic is exploding. Meanwhile, several neurolaw blogs have been launched on the Internet, and a growing number of law schools offer courses in neuroscience and the law.[9]

Across the United States, prosecutors, defenders, and judges are educating themselves about the science of brain imaging through conferences and seminars—wisely so, now that brain-based testimony is commonplace in capital defenses. In fact, several convicted murderers have already appealed their death sentences on the grounds that their lawyers wrongly denied them brain-scan evaluations. "Lawyers and judges have grown up thinking that social science is soft," says constitutional law scholar David Faigman. "Neuroscience gives the courts a hook."[10]

The hook, of course, rests on the assumption that brain function, and brain images more specifically, can help explain the defendant's behavior. At first blush, this makes sense; after all, if the brain determines an offender's state of mind, then forensic experts should be able to probe his or her brain to help resolve questions of culpability. In reality, though, this is an exceedingly tall order. For the brain to take the stand in an eloquent manner, neuroscience must first be translated accurately into concepts that have meaning within the law.

The *Simmons* case raised a host of fundamental issues within neurolaw. The first set of concerns is technical: What, precisely, is the relationship between brain function, as represented by neuroimages, and criminal behavior? The second set of concerns is legal: What is the effect of neuroscientific evidence on triers of fact? As one can imagine, overstating the significance of scans in forensic settings can have dire consequences for the accused and the criminal justice system more broadly. The third set is conceptual and philosophical: How does the law regard causal explanations of behavior in determining guilt? How do most potential jurors understand—or misunderstand—the relationship between biological explanations of behavior and the capacity for self-control and, thus, criminal responsibility?

Nothing less than the legal system's authority to hold offenders accountable hinges on a clear grasp of the relationship of mental contents and mental capacity to ascriptions of responsibility. Let us be more specific. How can neuroscientific data assist the law in ascertaining guilt? To answer this question, we need to understand how the law determines guilt. Some brief background: American criminal law holds a person responsible for a crime if he or she intended to commit a prohibited act. This mental state is called *mens rea*, or guilty mind, which generally requires either intent or reck-lessness. Without evidence of *mens rea*, the law cannot hold a person criminally responsible. For example, hitting and killing a pedestrian when one's car goes out of control does not entail *mens rea*, whereas aiming one's car at a pedestrian, stepping on the gas, and running over that person does.

There are circumstances, however, under which a person may commit a prohibited act and yet be cleared of blame. For example, self-defense permits a defendant to intentionally kill a wrongful ag-gressor who is threatening him or her with deadly force. This is con-sidered a "justification" for the defendant's action. In other situations, the defendant may be "excused," meaning that the defendant's ac-tions are still considered wrongful, but the defendant is deemed to be not responsible for them. Excuses include duress (if the defendant committed the crime "with a gun to his head," as the saying goes) and legal insanity.

The federal insanity-defense statute holds that a defendant can be excused if, "as a result of a severe mental disease or defect, [he] was unable to appreciate the nature and quality or the wrongfulness of his acts."[11] In essence, the defendant's mind was so deformed by mental illness that he or she was incapable of understanding the nature of the act and was incapable of conceiving common notions of right and wrong. Some states allow an insanity defense in which the defendant is said to have been unable to resist his or her impulses.

Let's turn now to the relationship between cause and excuse. As we saw, the law's concept of the person is that of an agent who is capable of acting at will and offering reasons for his or her actions. Rationality is the hallmark of responsibility. There may be myriad reasons why people commit crimes, but no matter the explanation—from bad neurons to bad parents to bad stars—defendants will be found legally blameworthy as long as their rational capacity remains largely intact. Obviously, if causation alone served to excuse behavior, then all behavior would need to be excused, and no one could ever be held responsible for his or her actions. Biological causes are accorded no special weight in the eyes of the law, even though many people harbor the intuition that they should wield more clout as an excusing condition. To fall into the trap of thinking that a biological cause means exoneration is to commit what legal scholar Stephen Morse calls the "fundamental psycho-legal error." As Morse notes, the law cares only whether a causal factor, no matter its nature, produced impairment so substantial as to deprive people of their rationality.[12]

IF brain scans are to play a scientifically legitimate role in determining criminal responsibility or in reducing a defendant's sentence, they need to be able to assist us in answering legal questions. That means, at bottom, that they must be amenable to being deciphered in such a way that they bear narrowly on potentially excusing or mitigating mental states, such as damaged capacity for reason or an impaired ability to form intent or exert self-control. As we will discover, the extent to which brain scans provide such guidance is more limited than many people realize.

In the case of Brian Dugan, an Illinois man facing the death penalty for the kidnapping, rape, and murder of ten-year-old Jeanine Nicarico, the defense team called on fMRI to show that his ability to distinguish right from wrong was profoundly impaired. Dugan was already serving two life sentences for other rapes and murders he had committed in Chicago. In 2009, when Dugan was fifty-two, his lawyers invoked fMRI evidence in the penalty phase of his trial to show that he was a psychopath, a morally disabled man whose sickness was such that he could not *feel* right from wrong or that he did not care about the distinction.[13]

As a psychopath, then, Dugan would have known that killing and raping Nicarico was against the law, but he would not have appreciated the moral gravity of these acts. This is not because psychopaths are entirely emotionless. They can feel great anger when they are insulted, humiliated, or rejected. And they can be masters of manipulation, which suggests that they may sometimes be good at reading certain emotions in others. But they tend to be extremely poor at empathizing emotionally and typically regard the pain or misery they've inflicted on others as being "their problem, not mine." Psychopaths also manifest a diminished capacity to learn from negative consequences that would ordinarily dampen aggressive impulses.

Psychologists typically quantify psychopathy by measuring three sets of traits: interpersonal deficits (such as grandiosity, arrogance, and deceitfulness); affective deficits (incapacity for love, guilt, or remorse); and impulsive and irresponsible behaviors. Experts estimate that psychopaths (the majority of whom are nonviolent) represent between 15 and 25 percent of the prison population and 1 percent of all people in the general population, with the rates being higher in men than women.[14]

Psychologist Kent A. Kiehl of the University of New Mexico was the star expert witness for Dugan. His job was to verify that Dugan met the diagnostic picture of a psychopath. Kiehl began by administering a standard, detailed interview known as the Hare Psychopathy Checklist—Revised, on which the defendant scored a stratospheric 38.5 out of a maximum score of 40. Kiehl also scanned Dugan's brain, guided by work suggesting that psychopaths suffer from deficits in emotional responding during moral decision making. Such deficits are in turn linked to impairments in brain regions that register feelings and assign emotional value to expectations and experiences.[15]

Kiehl tested Dugan using fMRI the same way he had tested over 1,000 inmates before him in research investigations. In those studies, Kiehl and his colleagues scanned psychopathic and nonpsychopathic prisoners (those with Hare scores under 30) as they reacted to three types of pictures: moral, nonmoral, and neutral. Examples of moral images included photographs of Klansmen and a burning cross, an adult screaming at a cowering child, and a person suffering a beating. The nonmoral images, such as a crying child, a vicious dog, and a gruesome facial tumor, were disturbing but lacked perpetrators. The neutral set consisted of pictures of people chatting, painting, and playing sports.[16]

Next, Kiehl's team asked the inmates to concentrate on the photos showing a moral violation in progress. As nonpsychopaths viewed the photos, their brains showed greater activity in what Kiehl calls the "paralimbic system" (an interconnected set of emotion-processing structures, including the anterior temporal cortex and the ventromedial prefrontal cortex) than when they looked at nonmoral and neutral pictures. In marked contrast, the brains of psychopathic subjects manifested similar low levels of activation in response to all three sets of images. When Kiehl tested Dugan's brain, he found the same basic psychopathic pattern. At trial, however, the judge did not allow Kiehl to show the scans of the defendant's abnormal paralimbic activity; he worried that they might confuse the jury. As a compromise, he did permit Kiehl to show jurors a diagram of the findings and to explain their meaning. In the end, the jury was not persuaded and sentenced Dugan to death.[17]

KIEHL's work is a recent chapter in the search for the source of criminality in the brain. As we've already seen, nineteenth-century phrenologists believed that bad behavior was rooted in bad character, which, in turn, stemmed from defective brain organization as reflected in the shape of the skull. Franz Joseph Gall, the father of phrenology, identified several brain "organs" that purportedly gave rise to criminal behavior if they were hypertrophied or atrophied. There was an organ of Murder—later renamed the organ of Destructiveness—and organs for Combativeness, Covetousness, and Secretiveness, all manifested as bumps in specific places on the skull. Phrenology, sometimes jokingly called "bumpology," exerted a strong influence on criminal law in both the United States and Europe in the early

and mid-1800s. Practitioners routinely testified in support of reduced punishment for the convicted. They also assisted judges in
determining whether murderers were insane or capable of planning
their crimes, and whether witnesses were reliable.[18]

In the waning days of penal phrenology, an Italian physician
named Cesare Lombroso advanced the idea that vicious crimes were
caused, not chosen. When he conducted a postmortem exam on a
serial rapist and murderer, he found an anomalous depression inside
the skull toward the rear midline region, where the cerebellum
would have resided. This hollow, he wrote, seemed like that found in
"the lower types of apes, rodents, and birds." In 1876, Lombroso
published *Criminal Man*, in which he advanced the idea that lifelong
violent criminals are atavistic throwbacks to savages. "Theoretical
ethics passes over these diseased brains as oil does over marble,
without penetrating it," he wrote. These congenital offenders
required permanent isolation for the safety of all, whereas other,
more biologically evolved wrongdoers were to be educated and
reformed.[19]

Throughout the twentieth century, biological models of crime
jostled with psychoanalytic and sociological theories. The latter
were dominant, attributing chronic offending to psychological, economic, and political factors; social learning theory—the idea that
crime is an adopted behavior—was also influential. But biological
determinism saw a minor revival in the wake of the Detroit race riots in the summer of 1967. Neurosurgeons Vernon H. Mark and
William H. Sweet joined psychiatrist Frank R. Ervin to publish a
letter in the *Journal of the American Medical Association* titled "The
Role of Brain Disease in Riots and Urban Violence." Mark and Ervin
expanded on their views in a controversial 1970 book, *Violence and
the Brain*, arguing that violence was "related to brain malfunction"
and advocating treatment by introducing an electrode into a small
part of the limbic system to correct it. Their arguments caught the
attention of a handful of neurosurgeons and some prison adminis-

trators, as well as the U.S. Department of Justice. Although only a few surgeries were actually performed on inmates, public concern about "identity-destroying" and inhumane treatment of prisoners mounted. At a congressional hearing in 1973, the director of the National Institute of Mental Health testified that surgery should not be used to change behavior in nonpsychiatric patients.[20]

NEUROSCIENCE may one day contribute to determinations of capacity for reason and impulse control, but a plethora of technical issues stand in the way. For one thing—and this is a glaring caveat—by the time brain scans are performed, the deed has already been done. Brains change over the years; they age, and they reorganize through injury and experience. Only rarely can brain scans be said to depict the neural correlates of a defendant's mental state at the time of the crime, as, for example, when they show long-standing defects that presumably reflect stable features of the defendant's cognitive capacities. Even then, showing that these abnormalities predate the crime is more easily said than done.

Theoretically, then, it is possible that the abnormalities in some of the emotion-mediating structures in Dugan's brain preceded and contributed to his crime, which he committed over two decades ago. But they also might have arisen as a consequence of Dugan's having spent decades in prison. Alternatively, the abnormalities may have been a sheer coincidence, having no direct bearing on his crime at all. Ideally, we would want to know whether every person with that pattern is a murderer, as well as whether everyone without the pattern is a nonmurderer. But that standard of causality is unrealistically high.[21]

This is not to say that brain-based techniques will not one day contribute uniquely to the facts of a case. This will depend on whether they can detect abnormalities tightly linked to cognitive deficits relevant to the defendant's competence to appreciate the wrongfulness of an act, form intent, learn the basic rules of the law, and conform to its demands. In Dugan's case, for example, the assumption behind measuring reactions to pictures of moral violations was that his reactions would reveal characteristics of brain processing that are closely associated with the constellation of thoughts and emotions that led Dugan to stalk, rape, and kill Nicarico. This link is suggestive at best and has not been demonstrated to even a modest level of certainty.

For now, neurologists, psychiatrists, and psychologists do not know, except in the most extreme instances of brain damage or acute injury, whether a given brain abnormality is relevant to the criminal behavior in question.[22] There are many reasons for this uncertainty.

Although brain imaging can measure fluctuations in blood oxygen, as we've seen, interpreting changes in brain activation as evidence that a defendant fails to meet legal criteria for full responsibility—significantly impaired rationality, inability to form intent, weakened impulse control, and so on—is not yet on a scientifically firm footing. It is important, too, that "abnormalities" do not necessarily have functional significance. Neurologists have recognized for decades that many people with "bad" brains (showing suspicious lesions or, in the case of functional scans, abnormal activation patterns) are law abiding. Frontal-lobe damage, for example, is statistically associated with increased aggression, yet most people with such damage are not hostile or violent. Presumably, the extensive connectivity across brain regions allows some areas to modulate and compensate for others. Conversely, some individuals with severe behavioral problems display few or no defects when their brains are imaged.[23]

An example of an impressive-looking, though ultimately irrelevant, brain defect turned up in Herbert Weinstein, whose case is now a classic in neurolaw.[24] In 1991, the sixty-five-year-old retired New York advertising executive strangled his wife in the course of an argument and pushed her out the bedroom window of their twelfth-floor apartment, hoping to make the murder look like a suicide. The police caught him as he tried to sneak out the back of his Upper East Side building. Weinstein was charged with murder in the second degree, and his lawyers set about preparing an insanity defense by sending their client for a neurological assessment, including a PET scan. Imaging was rarely used in forensic settings at the time, and the lawyers viewed it as a Hail Mary attempt to turn up anything irregular.

The scan turned out to be striking. Prominently visible in Weinstein's left frontal lobe was a gaping black void the size of a quail egg. It was a fluid-filled cyst that had formed within the weblike tissues surrounding the brain. Over many years, the cyst had expanded slowly into the underside of the frontal lobe, displacing and compressing brain tissue that in the scan glowed red and green, colors representing regions of "hypometabolism," or decreased energy use. According to the defense, this anomaly produced a severe impairment in Weinstein's ability to appreciate the difference between right and wrong.[25]

Despite the dramatic visuals, most radiologists who examined the evidence concluded that Weinstein's cyst had little effect on the functioning of his brain. "The PET was abnormal," said a psychiatrist who served as a witness for the prosecution, "[but] that has nothing to do with the fact he threw his wife out the window."[26] After the judge decided to admit some of the neurological evidence at trial, prosecutors agreed to let Weinstein plead guilty to manslaughter. Many legal scholars credit the PET scan for winning him a reduced sentence of only seven years.

Another reason we must be cautious is that an apparent brain abnormality may not turn out to be a true abnormality. When researchers examine a defendant's brain scan, they compare it with a control brain scan, a composite of "normal" brains created from the aggregate data of many averaged subjects. What jurors might not realize is that as a result of the considerable variability among people's brains, the defendant's pattern of brain activation might well resemble that of some of the normal subjects.[27] By way of analogy, consider that although the average American man is five feet, eight inches tall, weighs 175 pounds, and is right-handed, brown-eyed, Caucasian, and forty years old, very few American men share all those exact traits. Thus the pooling process can make a defendant's brain appear defective when it is really just a variant of normal.

Even when there is a clear-cut relationship between brain defect and dangerous impulse, how can we know whether the defendant was genuinely helpless to resist? Consider the intriguing case of a forty-year-old schoolteacher who developed a strong interest in child pornography as the result of a brain tumor. Earlier in his life, he had been interested in adult pornography, but in 2000, for the first time ever, he reportedly made overt sexual advances to his young step-daughter and to adult women. Around that time, he underwent an MRI to diagnose neurological problems, such as headaches, an abnormal gait, and an inability to write words. The MRI showed that his right orbitofrontal cortex had been invaded by a large tumor. Doctors excised the growth, and the patient said that his pedophilic urges had disappeared completely. Yet a year later, his appetite for child pornography returned. Sure enough, a brain scan demonstrated that the tumor had grown back.[28]

The tumor was almost certainly a cause of the teacher's intense sexual appetite; at the very least, it probably released a brake on a preexisting desire. In any case, not everyone who experiences an urge acts on it. In fact, shortly before the tumor was discovered, the schoolteacher went to the emergency room, complaining of a strong desire to rape his landlady in addition to his headache and other neurological symptoms. Possibly, the urge to rape frightened him so much that he sought refuge to protect both himself and the landlady.

Neuroscience cannot yet distinguish those who *could not* control themselves from those who *did not* control themselves, nor from those in between who managed to wrestle their impulses to the ground. Perhaps neuroscience will never succeed in making these distinctions. A great deal more needs to be known about the nature of control systems in the brain and how they interact with the circuitry of motivation and desire. Scientists must be able to show that specific imaged patterns correlate tightly with the kinds of deficits in reasoning and self-control that constitute excusing or mitigating conditions before brain scans can become a sound source of forensic evidence.

Brain scans can also be remarkably silent in some of the most extreme cases. Consider Andrea Yates. The case of the thirty-six-year-old Houston mother who killed her five young children in 2001 is a wrenching example of a rationality defect that eventually led to acquittal by reason of insanity. One June morning, after her husband had left for work, Yates methodically drowned her four boys and infant daughter in a bathtub. She then called the police, saying that she needed an ambulance. "I just killed my kids," she said to the officer who came to her door. Yates had been struggling with postpartum psychosis following the birth of her daughter seven months earlier. As she told psychiatrists who interviewed her in jail, her children "were doomed to perish in the fires of Hell" and "had to die to be saved."[29] Mental illness had so profoundly corrupted Yates's ability to get the facts right that she thought that killing her children would save them from eternal torment.

At trial, the jury rejected Yates's insanity plea on the strict basis that she knew that murder was wrong. (Under Texas law, the insanity standard hinges narrowly on whether the defendant knew that his or her behavior was legally wrong. Even individuals clearly suffering from severe mental illness, as was Yates, can fail to meet this standard.) Several years later, however, in a memorable turn of events, an appeals court overturned her conviction because it became clear that prosecutors had used incorrect testimony to suggest that she had gotten the idea for the killings from a nonexistent episode of the television series *Law and Order*. At retrial in 2006,

Yates's lawyers presented their original arguments, and a jury found her not guilty by reason of insanity.[30]

Notably, Yates's lawyers did not present brain-scan evidence. But even had they done so, a brain scan would not have revealed signs of her illness. In fact, many years after Yates's trials, brain scans still cannot prove that a woman has postpartum psychosis, let alone that she did not grasp the significance of a crime she committed while acutely ill. This state of affairs may one day change, however, as imaging technology evolves and if it succeeds in delineating new diagnostic categories based on brain dysfunction.[31] But such categories are nowhere on the forseeable horizon.

NEUROSCIENCE evidence is becoming ever more popular in the courtroom, according to neurolaw experts. Between 2005 and 2009, the number of criminal cases in which neurological or behavioral genetic evidence was introduced doubled. In capital cases, judges have wide latitude in allowing defense teams to present evidence that sheds doubt on moral culpability for sentencing purposes. In cases in which legal culpability is on the line, however, the threshold for admitting evidence must be based on a higher standard of sound science.[32]

The first time Christopher Plourd, a San Diego–based criminal defense attorney, used PET-scan evidence in a murder trial, he was impressed by its persuasiveness. "Here was this nice color image we could enlarge, which the medical expert could point to," he told a journalist about the trial, which took place in the early 1990s. "It documented that this guy had a rotten spot in his brain. The jury glommed onto that."[33]

Lawyers have glommed onto it, too. "A mind in turmoil can be portrayed with scientific precision, and that picture can help humanize the accused and enlighten decision-makers on the limits of liability and punishment," said Ken Strutin, director of legal information services at the New York State Defenders' Association. Many legal scholars and neuroscientists are troubled by such overblown rhetoric—and for good reason. If people can be seduced by the aura of scientific authority surrounding brain images, might the scans lead jurors to believe that they are actually observing an irresistible biological cause of criminality? Could such misplaced faith in neurorealism distract jurors from salient but more mundane forms of information contained in conventional kinds of evidence?[34]

The legal term for this kind of distortion is "prejudice." To be clear, prejudice in this context has nothing to do with attitudes toward the defendant's race or ethnicity, but rather involves attitudes toward the evidence. Judges must be alert to the possibility that jurors might attribute far greater precision and validity to a piece of evidence than is otherwise appropriate. Admittedly, the evidence isn't entirely consistent, but a few studies appear to bolster the expectation that when brain images accompany explanations of behavior, people find those explanations more compelling. For example, psychologist Madeleine Keehner and colleagues showed readers scientific reports accompanied by brain-scan images of an increasingly detailed nature. The more concrete, three-dimensional, and "brain-like" the image, the more likely it was to persuade naive readers than was an accompanying written report that contained good scientific reasoning.[35]

In a much-cited experiment, psychologists David P. McCabe and Alan D. Castel presented college students with flawed explanations of mental phenomena paired with a brain scan. By deliberately making the explanations illogical, the researchers could acquire a better grasp of the potential of images to distort the meaning of an explanatory narrative. They told subjects that people can improve their mathematical ability by watching television and tried to persuade them—nonsensically—that data depicting only a statistical association between television viewing and math ability would count as proof for this claim (such an inference, of course, risks confusing correlation with causation).

McCabe and Castel divided their subjects into three groups and displayed to each group the bogus data with a different accompanying explanation. One group received a written explanation of the findings; the second received the explanation along with a bar graph measuring temporal-lobe activity; and the third received the study description along with a multicolored brain image. Participants rated the soundness of the reasoning in the math-and-TV vignette as more believable when it was accompanied by the brain images. Along similar lines, when psychologist Deena Weisberg and colleagues inserted the phrase "brain scans show" into illogical explanations of behavior, those explanations became more compelling to neuroscience nonexperts (but not neuroscience experts). Taken together, these findings raise the possibility that neuroimagery—sometimes humorously called "brain porn"—and neurolanguage can seduce jurors and others into drawing erroneous conclusions.[36]

To limit the introduction of misleading evidence into trial, Federal Rule of Evidence 403 directs judges to weigh the potential of expert testimony and exhibits to prejudice jurors against their probative value—their potential to help the jurors resolve the legal issue in question.[37] Judges can refuse to admit brain-scan evidence into trial if they believe that it will unduly bias the jury against the defendant, as did the judge in the case of Dugan, the probable psychopath who kidnapped and killed a young girl.

It is probably impossible to know whether brain scans have prejudiced jurors in any given case. Short of gaining access to transcripts of jury deliberations or performing exit interviews of jurors, how could researchers gauge jurors' interpretation of the evidence and the relative weight they accorded the array of courtroom information, from testimonies of expert witnesses to comportment of the defendant, attorney summations, or expression of remorse by the defendant? As an imperfect but helpful alternative, researchers have attempted to measure the influence of brain-based evidence on decision making by subjects acting as jurors.

Psychologist Michael Saks and colleagues sought to tease apart the myriad dimensions of brain-based evidence to gauge their influence on juror decision making. They recruited a large sample of subjects to read about a true-life robbery that turned into a gruesome murder. The mock jurors were split into several groups, each of which received a different explanation of why the defendant had been unable to form the intent to commit the murder. One group, for example, read expert testimony by a neuroscientist describing the discovery of damage to the left frontal lobe on a scan. A second group viewed a picture of the actual scan with the defect visible. Another group read the testimony of a psychologist diagnosing the defendant with a personality disorder.[38]

In the end, jurors told that the defendant suffered from a personality disorder, rather than a specific brain defect, penalized the offender more severely. All brain-based explanations, however, were weighted equally. Only when subjects learned that the offender faced execution—as they did in a related study by Saks's team—did the brain-scan evidence lead to the highest rate of mitigation to a life sentence. Other evidence that pointed to a genetic predisposition to violence or evidence of neurological defect on physical examination was not as compelling. Saks speculated that brain images are most influential when the ultimate punishment—death—is at hand.[39]

What do these findings mean? Can mock-juror judgments made in isolation about abstract cases come close to the kinds of decisions that actual jurors make in a courtroom? After all, when jurors hear a real case, there is a lot of detail they need to integrate: they absorb from a raft of witnesses, watch cross-examination of experts, hear closing statements from lawyers and the judge's instructions, and engage in lengthy discussions with fellow jurors. Most powerfully, perhaps, they also know that their decision affects the freedom and often the very lives of real people.[40]

Creative researchers can get around some of these hurdles by using real jurors from a juror pool, reenacting actual cross-examinations of experts and closing arguments, asking for verdicts before and after deliberation, and testing jurors about relevant facts to see whether brain-scan images confused them or distracted them from crucial testimony.[41] The "life-in-the-balance" element of capital trials, however, would be very hard and arguably impossible to reproduce in a research setting.

In actual cases in which the defense introduced brain-scan evidence, the effects have been mixed. In some instances, presentations of scans appeared to help the defense win lesser sentences or exculpation for their clients, but in others they exerted little apparent influence. But there is little doubt, as we'll soon see, that neurobiological explanations, as opposed to psychological or social ones, give rise to impressively different views of moral responsibility. Pleading that "my brain made me do it" weakens ascriptions of responsibility in a way that "my lousy childhood made me do it" does not. In the first case, neurological terms suggest rigid internal processes that lead inexorably to only one behavior. But when a theory of behavior is expressed in psychological terms, it is easier to imagine a person acting differently—a cognitive bias that brain scans may reinforce.[42]

Psychologist John Monterosso and colleagues discovered in 2005 that giving subjects physiological explanations, such as a "chemical imbalance," for crimes like arson and murder resulted in greater rates of exoneration than did psychological explanations, such as childhood abuse. Psychologists Jessica Gurley and David Marcus found that expert testimony that included either photos of the brain scan or a narrative about how the brain damage occurred led to acquittal by roughly one-third of subjects, a significantly higher rate than that for acquittal testimony devoid of neurological evidence.

Likewise, in 2003, psychologist Wendy P. Heath and colleagues investigated the effects of a wide variety of explanations for criminal behavior (including biological, psychological, and environmental accounts). They reported that subjects rated the biological ones as more credible and deemed the wrongdoer less culpable. Finally, in 2012, University of Utah researchers asked actual trial judges to review a fictional case in which a young psychopathic man savagely beat the manager of a food restaurant. Some of the judges read testimony from a neurobiologist who tested the defendant and discovered that he possessed a gene variant associated with violent behavior and callous disregard for the suffering of others. Those judges who read the neurobiologist's testimony handed down an average sentence of thirteen years—a full year less than the average sentence issued by the judges who had not seen the testimony about genetics and violence.[43]

In light of these findings, it is easy to understand why juvenile justice advocates were galvanized by the Supreme Court's *Simmons* decision. Although the word "brain" does not appear in any of the justices' written opinions—the majority opinion makes clear that their decision was based on "evolving standards of decency that mark the progress of a maturing society"—reformers hailed the decision as a vindication of their years-long effort to deploy brain science to reduce the length of juvenile sentences and to incarcerate violent teen offenders in forensic rehabilitation centers rather than adult penal institutions. As one reformer put it, the new "hard science" of brain imaging should compel the legal system to consider adolescents to be in a "natural state" of diminished capacity.[44]

And they've made inroads. In a 2010 case, *Graham v. Florida*, the U.S. Supreme Court banned sentences of life without parole for juveniles convicted of nonlethal crimes to permit offenders "a chance to demonstrate growth and maturity." Although the case did not rely heavily on neuroscience, Justice Anthony Kennedy, writing for the majority, mentioned it, noting that the "biological basis for differences in juvenile conduct provides further support for the conclusion that less culpability should attach to juvenile conduct than to similar conduct by adults." That same statement was invoked in the majority opinion in *Miller v. Alabama*, a 2012 Supreme Court ruling that a mandatory sentence of life without parole for juvenile killers violates the constitutional protection against cruel and unusual punishment. At the state level, the California legislature passed a bill in 2012 that allows some juveniles serving life without parole to earn parole after serving twenty-five years. "The neuroscience is clear. . . . [Teens'] impulse control, planning, and critical thinking skills are not fully developed," said the sponsoring senator, who is also a child psychologist.[45]

Against this political backdrop are some inconvenient realities that argue against invoking the teen brain to explain the behavior of violent youth in one broad brush stroke. For one thing, neuroscience adds little to what every parent already knows. Teens, especially boys, can be reckless. They tend to drive too fast, drink too much, and skateboard down staircases. But in the case of Christopher Simmons, how strongly did the immaturity of his frontal-lobe and amygdala complex explain his actions? His crime, after all, was not impulsive; Simmons had a lethal plan before he broke into Crook's home. Nor did he need to possess a fully formed brain to know that throwing a person off a bridge is wrong. An average nine-year-old grasps the finality of death.[46]

In truth, there is great variation in teen behavior. In part, this is because brains are bathed in culture and circumstance. Consider a teen with working parents who must care for younger siblings. Growing up fast because of life experience or in response to demands made by others can result in well-honed capacities for judgment and self-discipline. Thus the teen brain develops in and is molded by a dynamic environment. The vast majority of teens who harbor fantasies of violence do not act on them. Rates of teen violence and murder vary markedly from country to country. Juvenile delinquency in some preindustrial communities has increased over the course of a mere generation or two after the introduction of Western influences, such as television.[17] The lesson is this: Although the neuroscience of the adolescent brain helps us construct a plausible biological account of why adolescents can be more impetuous than adults, it says little about any individual teen offender.

Granted, it may seem harsh to fault juvenile justice advocates for exaggerating the degree to which adolescent brain development dictates their behavior—and in their telling, unsurprisingly, the degree is virtually 100 percent. But good intentions aside, these reformers would do well to remember that neuroscientific evidence is a knife that cuts both ways: If teens' brains render them irresponsible, what are the implications for the rights or opportunities teens now enjoy? Are they too immature to enter into contracts, as one state senator insists? To get an abortion, as pro-life advocates claim? To play violent video games, as consumer watchdogs allege?[48]

Biological explanations can influence the fate of adult defendants as well. Some jurors might think that defective self-control, decision making, or reasoning warrant longer, not shorter, sentences. On occasion, state prosecutors have introduced neurological evidence to stoke the perception that a defendant is destined for future violence and is therefore too dangerous to return to the community. In one instance, a defense team argued that a client's inborn genetic predisposition to crime warranted mercy, but prosecutors successfully exploited the genetic evidence to secure a harsh sentence based on the premise that an inborn proclivity to violence made him more of a public threat. Parole boards could use similar logic to deny applications for release.[49]

Finally, should the defense prevail and brain-based evidence result in more lenient sentences, the public might demand that criminals be contained beyond their completed sentences out of fear that they will commit new violent crimes. Current law surrounding sexually violent predators is a model here. In the name of public safety, many states detain sexually violent predators well beyond the completion of their formal sentence if a court determines that they have a high risk of reoffending.[50] In the end, preventive detention based on brain-derived evidence is a highly fraught prospect. From a scientific angle, the accuracy with which experts can predict long-term future dangerousness is still quite poor, although some neuroscientists are betting that brain science can improve the odds. From a civil liberties standpoint, the classic struggle over how best to weigh community safety against another's individual liberty is invoked anew.

THE jury is still out, so to speak, on the impact of neuroscience on the practice of law. In capital cases, judges are confronting brain-scan evidence more and more, but the swirl of factors that influence jurors' decisions makes it difficult for investigators to parse and evaluate its impact. And even in instances in which images are taken into account and influence a juror's decision, they might not lead to a "wrong verdict." Perhaps acquittal by reason of insanity or mitigation is indeed the just outcome in certain cases. As for the value of functional brain images in the courtroom, the picture is clearer: Scans are not yet able to go beyond the insights that lawyers and experts can glean through traditional forensic methods. At best, they run afoul of the pesky neuroredundancy problem by looking to images for answers they can obtain by other means.

Few doubt that a few people lack the capacity to understand the law, and that for others it is hard to suppress impulses—perhaps close to impossible in certain instances. Nevertheless, brain scans cannot yet tell us who those people are. The tragic case of Andrea Yates is a powerful illustration. Something had gone terribly wrong in her brain. Yet no scan or other biological test could have illuminated her crippled state of mind better than the clinicians who talked to Yates and her relatives and read her psychiatric records. To understand Andrea Yates and her crime, "you need to understand why," says forensic psychiatrist Phillip Resnick, who testified on Yates's behalf. "And you can't see why on an fMRI."[51]

In our view, the potential for functional brain imaging to mislead currently exceeds its capacity to inform, although the ratio may eventually shift in favor of the value of scans for some purposes as technical advances emerge. But until neuroscientists and legal experts become able to translate information about brain function into the legal requirements for criminal responsibility, lawyers, jurors, and judges will still need to rely on traditional methods of assessing the defendant: interviews, observations, witness reports, psychiatric history, and well-established clinical assessments.[52] It is from these methods, in any case, that a subtler appreciation of the defendant's mental state can be inferred.

Brain scans can never fully capture the criminal mind, or any mind for that matter, but perhaps one day they will be better able to identify neural patterns tightly linked to profound derangements in rational capacity and self-control. Also welcome would be neuroscientific guidance in vexing problems, such as identifying defendants who are faking mental illness to avoid standing trial, or distinguishing false memories of sexual abuse from accurate ones.

Whether the formidable technical hurdles involved in drawing inferences from imaging can be cleared remains to be seen, but even if they are, subjective judgments are inescapable. Let us say, for example, that brain evidence will someday be able to show that a defendant lacks the capacity to act rationally. Society will still need to grapple with the question of just how much capacity a defendant must have for jurors to deem him or her not responsible or less culpable for a crime. Where should experts draw the line? How much prefrontal abnormality, unfinished myelination, or overdriving limbic activity is necessary to support the claim that a defendant could not have exerted self-control, "felt" no difference between right and wrong, or was unable to reason cogently?[53]

In the case of psychopaths, asks legal scholar Ken Levy, "Should they be held criminally culpable if they rationally know the difference between right and wrong but can't emotionally grasp the moral gravity of their criminal actions?"[54] As for teens' eligibility for the death penalty, is there anything magical about the age of eighteen? For better or worse, since 2005 the law has drawn the line there, but there is no precise neurodevelopmental threshold at which a young person morphs from a hotheaded teen into a measured decision maker. Maturation unfolds along a continuum that varies widely among individuals and depends, in part, on familial, social, and cultural settings that no child can control. There are compelling ethical reasons for eliminating the death penalty for juvenile killers. But whether the neurobiology of their collective brains should categorically exclude teens from certain forms of punishment is a question that science alone cannot answer.

In the next chapter, we explore a knotty philosophical question raised by neuroscientific progress: Does brain science threaten the notion that people—all people, not just criminals—possess the freedom to act?

A growing number of scientists, citing the explosion in knowledge about the brain, are now challenging the law's bedrock assumption that, with some exceptions, people are rational, choosing, responsible creatures.[55] Their argument goes like this: Given that our conduct is caused by brain function, which, in turn, is caused by the interplay of genes and environment (factors over which we have no genuine control), we do not truly "choose" our actions. As a result, we cannot be held morally responsible for any wrongdoing. Clearly, this perspective holds significant implications for the design of our criminal justice system.

"Progress in understanding the chemical basis of behavior will make it increasingly untenable to retain a belief in the concept of free will," writes biologist Anthony R. Cashmore.[56] "I propose that the time is opportune for society to reevaluate our thinking concerning the concept of free will, as well as the policies of the criminal justice system." Yet this is more easily said than done. Are we even capable of giving up on—or substantially modifying—our intuitive view of free will? And can neuroscience provide a convincing case for doing so? In the next chapter, we'll explain why we doubt that it can.

6

THE FUTURE OF BLAME

Neuroscience and Moral Responsibility

I N MAY 1924, two young men set out to kidnap and murder a child of an affluent family. Nathan F. Leopold Jr., aged nineteen, and Richard Loeb, eighteen, had spent months planning and rehearsing what they called "the perfect crime." On the day of the killing, they chose a convenient victim: fourteen-year-old Bobby Franks, who was Loeb's second cousin and the son of a local millionaire. As Franks walked home from school on a late afternoon in Chicago's leafy Hyde Park neighborhood, the two pulled up alongside him in a rented roadster and invited him to hop in. After a few minutes of small talk about a tennis racket, they bludgeoned the boy to death and drove to the outskirts of a town near Indiana. Once there, they doused Franks's face with hydrochloric acid to hinder identification by the police and hurriedly stashed his naked corpse in a drainpipe.[1]

Later that evening, the killers were back in Leopold's elegant Hyde Park home. They drank and played cards, interrupting their game around midnight to phone Franks's family and tell them to expect a ransom note for their kidnapped son. Leopold and Loeb never fathomed that they might be caught. These brilliant sons of privileged Chicago families—Leopold allegedly had an IQ of 200;

Loeb had graduated from college by age eighteen—believed themselves exempt from the laws that governed ordinary men.

Several days later, their plan unraveled when police found a distinctive pair of horn-rimmed eyeglasses at the crime scene and traced them to Leopold. Shortly thereafter, the two were indicted on charges of kidnapping and murder. Their parents hired famed attorney Clarence Darrow to defend them for committing what came to be known as "the crime of the century."[2]

The monthlong trial culminated in August 1924 with a bravura summation by Clarence Darrow arguing for life in prison for the two instead of death by hanging:

> Why did they kill little Bobby Franks? Not for money, not for spite; not for hate. . . . They killed him because they were made that way. Because somewhere in the infinite processes that go to the making up of the boy or the man something slipped, and those unfortunate lads sit here hated, despised, outcasts, with the community shouting for their blood.[3]

Leopold and Loeb's actions, in Darrow's telling, were just part of the natural order of the world: "Nature is strong and she is pitiless. . . . We are her victims," Darrow intoned. "Each act, criminal or otherwise, follows a cause; [and] given the same conditions, the same result will follow forever and ever."[4]

In the end, the judge spared Leopold and Loeb the gallows, sentencing each to life in prison for murder plus ninety-nine years for kidnapping—not because they were victims of nature, an argument the judge explicitly rejected, but because of their youth.[5]

DARROW's plea was remarkable. If "each act follows a cause," then all of us, not just Leopold and Loeb, are nature's victims. Bold as it was, however, the claim was not original. It drew on the ancient philosophical doctrine known as determinism, which states that every event is completely caused, or determined, by what happened leading up to it. Our decisions are the inevitable product of a vast array of influences—our genes (and the evolutionary history they represent), the mechanisms of our brains, our upbringing, and the physical and social environments in which we live. These forces converge to produce one and only one specific act, be it "choosing" soup over salad or murder over mercy. To borrow Darrow's words, you have "no more power than a machine to escape the law of cause and effect."[6]

What would it mean to live in a world in which people are simply mechanical devices responding to natural laws beyond their control, bobbing like corks in a sea of causes? If determinism is true, then the consequences are profound. First, we would need to radically overhaul our conception of moral responsibility. After all, if the choice you make in a given situation is preordained—is the only choice you *can* make—then what are we to do about blame? Absent the capacity to choose, according to a school of thought called hard determinism, there cannot be any blame. And if no one can be blamed, no one is morally deserving of punishment. If you commit an evil deed, it is not your fault. Nor is it to your credit if you behave like a saint. This account of human agency is devastating to the idea of free will (or "ultimate" freedom, as some philosophers call it).[7]

Hard determinists believe that society should adjust its legal practices accordingly. Philosopher-neuroscientist Joshua Greene and psychologist Jonathan Cohen contend that neuroscience has a special role to play in giving these age-old arguments more rhetorical bite. "New neuroscience will affect the way we view the law, not by furnishing us with new ideas or arguments about the nature of human action, but by breathing new life into old ones," they write. "[It] can help us see that all behavior is mechanical, that all behavior is produced by chains of physical events that ultimately reach back to forces beyond the agent's control," Greene adds. For emphasis, he and Cohen invoke an old French proverb, *Tout comprendre, c'est tout pardonner* (To know all is to forgive all). Their ultimate hope is that society will discard blame-based punishment as a nasty relic of a pre-neuroscientific age and insert in its place penalties whose purpose is to shape future behavior.[8]

Evolutionary biologist Richard Dawkins elaborates on the notion of the criminal offender as something of a machine. He invokes the example of a car that has stopped working. "Instead of beating the car," he notes, "we would investigate the problem. Is the carburetor flooded? Are the sparking plugs or distributor points damp? Has it simply run out of gas? Why do we not react in the same way to a defective man: a murderer, say, or a rapist? . . . I fear it is unlikely that I shall ever reach that level of enlightenment." Biologist Robert Sapolsky extends the analogy. We do not ponder whether to forgive the car, he says; instead, we try to protect society from it. "Although it may seem dehumanizing to medicalize people into being broken cars, it can still be vastly more humane than moralizing them into being sinners." This reasoning echoes Darrow's appeal to the judge that Nathan Leopold and Richard Loeb were merely "two young men who should be examined in a psychopathic hospital and treated kindly and with care."[9]

Of course, people are not like cars or other inanimate, nonconscious entities. Cars do not respond to knowledge, sanctions, or rewards, but people do. And it is for this reason that they are capable of being ruled by law in the first place. Hard determinism does not dispute this point. It acknowledges that people are educable, that they are constantly assimilating new information and, therefore, always learning. Take the example of shifting social norms surrounding drunk driving and domestic violence. As people learned of stiffer penalties for these actions, more of them came to think of those acts as wrongful.[10] Warnings operate on beliefs about the likely results of one's actions. New information builds on old experience and current context to guide subsequent action. Endowed with self-awareness, people but not cars can influence the outcome of their causal chains by making a decision to change their diet, their work habits, and their future.

Thus, although hard determinists reject retributive justice, also known as "just-desert theory," they do not deny that punishment sometimes has useful practical consequences, such as decreasing the chance that the criminal will reoffend. "Our modern understanding of the brain suggests a different approach. Blameworthiness should be removed from the legal argot," writes David Eagleman. Although utilitarian punishment carries no moral condemnation whatsoever, it exerts a salutary effect by prompting the criminal to reform and by dissuading would-be lawbreakers who observe the adverse consequences they could face.[11] And, given hard determinists' strict aim of reducing crime, the aversion may very well need to be highly unpleasant if that is the only way to deter future wrongdoing.

This general framework has been in place for thousands of years, but now some hard determinists are offering a new twist on this ancient view of the relationship between cause and blame. They are hazarding the empirical prediction that neuroscience will expose retributive punishment as scientifically mistaken. They predict that as neuroscientific study gradually reveals the underlying causes of behavior, the average person will come to see that his or her general sense of being free is just an illusion.

THE idea that we all live in a moral vacuum is indeed a shock to the collective sense of what people are like. We think we cause our own actions, and that we are responsible for the consequences. To contemplate how we could be free within a determined universe is to confront something "dark, puzzling, and not a little terrifying," wrote the American man of letters H. L. Mencken. British philosopher Isaiah Berlin tried imagining what life would be like without the concept of moral agency. "The entire vocabulary of human relations would suffer radical change," he concluded. "Such expressions as 'I should not have done x,' 'How could you have chosen x?' and so on, indeed the entire language of the criticism and assessment of one's own and others' conduct, would undergo a sharp transformation."[12]

Is there a way to preserve moral responsibility in a world in which all events leading up to the moment of choice determine precisely what that "choice" will be? Philosophers call this baffling question the "problem of free will and determinism." It is one of the most famous conceptual impasses in philosophy. At issue, to be clear, is not whether the capacity to choose is necessary for moral responsibility—most philosophers, most neuroscientists, and just about everyone else concur that it is. The disagreement is over the *kind* of freedom of choice that is necessary. As we've seen, hard determinists claim that the only kind that counts is "ultimate" freedom (sometimes called "absolute freedom" or "metaphysical free will"). Biologist Jerry A. Coyne puts it this way: "If you could rerun the tape of your life up to the moment you make a choice, with every aspect of the universe configured identically, free will means that your choice could have been different." But because that choice would not have been different, Coyne counsels that we should discard "the idea of punishment as retribution, which rests on the false notion that people can choose to do wrong."[13]

One approach to the thorny prospect of abolishing moral agency is to reject materialism outright and postulate a disembodied spirit—a "ghost in the machine"—that somehow directs the action from outside the physical flow of events. Such a dualist arrangement must be taken on faith and, like the existence of a godlike entity, cannot be disproved by science. This is because scientific inquiry depends on observation of measurable events that take place in the natural world; its purpose is to illuminate cause-and-effect relationships and test predictions based on them. As part of the supernatural realm, immaterial souls and a transcendent God are not amenable to the tools of science. So this strategy is a scientific dead end.

Another alternative is to assert that our behavior is independent of preexisting forces. In such a "causal vacuum," people are liberated from their own preferences, attitudes, and beliefs.[11] As a result, they can take more than one possible course of action under a given set of circumstances. Philosophers call this doctrine "libertarianism" (no relation to the political orientation of the same name). But there is no salvation here. An account of behavior in which there is no source of agency whatsoever only ushers in another kind of existential migraine. Human behavior arising randomly and haphazardly out of thin air would still count as behavior that is beyond the actor's control. How could anyone be considered free under these conditions either?

Consider, now, a third possibility called "compatibilism," which holds that freedom and moral responsibility can coexist in a way that does not require either libertarianism or determinism to be false. The argument is as follows: Even if human beings lack ultimate freedom (that is, they lack the capacity to have done otherwise), we can consider mentally intact adults morally responsible because they have the abilities to engage in conscious deliberation, follow rules, and generally control themselves.[15]

Eighteenth-century philosopher David Hume provided an influential statement of compatibilism by emphasizing that an agent's actions are free if they are caused by his or her will and desires, even if the will and desires are causally determined. Granted, we may be no more free from nature's chain of cause and effect than trees and butterflies. We lack responsibility in the ultimate sense; a deterministic universe does not allow for it. But as long as an actor's values and beliefs are causally relevant to his or her actions, then moral agency exists in the "ordinary sense," as British philosopher Janet Radcliffe Richards calls it. In short, the freedom to do otherwise is not the kind of freedom required for moral responsibility. If people can step back from competing desires, make a reasoned decision among them, and act on the basis of that decision, they possess capacities sufficient for free will. It is how they act after having reflected on things that warrants praise or censure.[16]

The capacity to be responsible in the ordinary sense seems to comport with people's general intuition of what it means to be a moral agent. Psychologist Roy F. Baumeister and colleagues found that subjects judged actions as "free" when they involved exertion of self-control, rational choice, planning, and initiative. For the average person, then, "free will" entails the capacities to be guided by reason, to evaluate a complex situation, and to conform to moral norms. Furthermore, a number of research teams have found that subjects who accept that events are caused by earlier events (and are less likely to hold hypothetical offenders to account) are more likely to deem them responsible when subsequently presented with scenarios in which actors commit a heinous crime that triggers angry feelings. In short, these data suggest that the average person accommodates the view that human beings are both determined and responsible.[17]

These findings may hearten compatibilists, but factual truth is not a popularity contest. So what if most people think that they and others have "free will"? Thus we come to the heart of the matter: The question whether humans can live in a material world and yet be morally responsible is not empirically testable. It is not a scientific problem. It is a conceptual and ethical impasse that has bedeviled thinkers since antiquity and is still without a resolution. Rest assured, our goal here is not to attempt to solve it; indeed, it may well be unsolvable.

What we do want to establish here—and this is a crucial point—is that neuroscience has not resolved it either. Those who believe that the absence of ultimate free will means that moral responsibility is an incoherent notion and that society should therefore abolish blame have already staked out a philosophical claim. Simply amassing more data on the workings of the brain may strengthen their conviction that determinism is correct, but it won't make their argument against the coexistence of determinism and moral agency any stronger.

LEOPOLD and Loeb's "decision" to murder Bobby Franks may have been the only one they could have made that May afternoon in Hyde Park. Still, few of us see them as automatons devoid of conscious thoughts and emotions. We believe instead that they harbored desires and acted for reasons, twisted as those motives were. Granted, their desires were not dispositions that the two men chose to have; nor could they be expected to know everything about why they had them. But in the end, Leopold and Loeb rehearsed their plan and consciously carried out the unspeakable thing they wanted to do.

Could we be wrong about that? Is it possible that Leopold and Loeb really were automatons after all? What if their actions did not flow from their conscious intentions and desires, but, instead, those actions happened *to* them, bypassing their conscious awareness entirely? To take this possibility a step further, might our actions do end runs around conscious deliberation in all of us all the time? This is the startling new challenge that some neuroscientists are introducing to the existence of free will: the possible absence of all consciously directed action. Such a radically reductionist prospect may strike many as outlandish, yet some of today's most highly respected scientists contend that individuals' subjective mental states—their yearnings, beliefs, and plans—play absolutely no role in bringing about their actions.[18]

To support this claim, they point to a series of arresting experiments conducted in the 1980s by physiologist Benjamin Libet. In his lab at the University of California at San Francisco, Libet wired subjects to an EEG and asked them to choose a random moment to lift a finger or move their wrist. He instructed them to watch the second hand of a clock and report its exact position when they felt the urge to move. Libet then measured electrical activity in a region of the frontal lobe called the supplementary motor area, which is involved in the planning of movements. What he found was striking: Activity in the motor area could be detected some four hundred milliseconds before subjects were aware of their decision to wiggle their fingers. In other words, the subjects' conscious awareness of the intention to move occurred too late in the sequence to influence the action. Instead of presaging movement, then, experience of the will to move followed it.[19]

Libet himself did not consider these results a wholesale repudiation of consciousness in guiding behavior. He speculated that although awareness came into play late in the sequence of events, people still have the freedom to "veto" or suppress actions that were caused by implicit processes outside awareness. That is, some say, we may not possess free will, but we have "free won't."[20] Yet others have interpreted Libet's results more radically, as proof that the mind or the person—the entity we think of as ourselves—is not calling the shots. Psychologist Daniel M. Wegner is a proponent of this view. We are so certain that we are in the driver's seat, he says, because we yearn to feel "authorship" of our actions.[21]

In an illustrative experiment, Wegner asked subjects to stick pins in a voodoo doll in the presence of another person who played a "victim." He told the victims to annoy the "witch doctors" by arriving late for the experiment and treating them rudely. Another set of victims weren't instructed to anger their witch doctors. All victims feigned a headache after the voodoo ceremony, but it was the witch doctors provoked by their victims who tended to claim that their ministrations caused the headache.[22]

Wegner puts forth other vivid instances where conscious intention is not in command. One such example involves a phenomenon called "confabulation," wherein individuals construct explanations for why they undertook actions that were clearly caused by forces external to them. For example, during a brain operation, neurosurgeons can induce patients to move their hands by stimulating the relevant controlling area on the motor cortex. When they ask patients why they moved their hands (patients are awake during some kinds of brain surgery), they often confabulate a reason they sincerely believe, such as wanting to get the doctors' attention.[23]

Those emerging from hypnotic suggestion may do so too. In fact, we all confabulate from time to time. Psychologist Timothy Wilson has coined the term "adaptive unconscious" to denote the kind of effortless and automatic cognitive processing that is inaccessible to awareness but that underwrites much of our behavior. Do we really "know," for example, why we fall in love with Mary but not Jane, take an instant dislike to Joel but not David, or choose one profession over another? We weave explanatory narratives to justify certain yearnings and choices, but can we ever know the truth? We are "strangers to ourselves," as Wilson puts it, not quite sure how and why we do many things and especially confused when we end up sabotaging our best interests.[24]

But are we estranged all the time? Libet's work on the awareness of intention to move stands as a powerful reminder that many of our actions are not willed in the way we normally think. But the conclusion that all our behavior is automatic all the time—that we are essentially beings in whom consciousness plays no causal role in behavior—is highly contested.[25] It's also a rather chilling one: If our actions are not guided by reasons, then why do anything at all? Thankfully, there are logically sound alternatives to that conclusion. After all, the fact that our intentions to act sometimes bypass our conscious desire to do so does not mean that our behaviors always "happen" to us, especially when the personal consequences or legal stakes for certain behaviors are high.

Our brains are notoriously good at thinking—you are doing that right now as you read along. We incubate ideas, deliberate on them, and intend the actions to which they have led us. Through this process, the self-modifying, plastic brain "learns" from experience and then "reasons" differently the next time.[26] Conscious thinking allows us to advance long-term goals, play out different scenarios, and reflect on past events, especially when novel situations present themselves.

In short, it seems highly implausible that our mental states never influence our behavior. Although Libet's subjects might well have moved their fingers or wrists at the very moment they did so without explicitly intending it, the wiggle itself was embedded in a sequence of deliberate moves. The subjects decided to participate in the experiment, navigated their way to the lab, and followed the experimenter's directions about what to do. In fact, they planned their actions in much the way a professor might strategize about which academic journal would best showcase his or her paper advancing the controversial thesis that decision making always bypasses conscious mental states.[27]

In the end, activity is a combined product of the automatic and the analytic. Think about it. Many of us can touch-type without thinking, yet we painstakingly choose the words we type on our job applications, dating profiles, and prenuptial agreements. Or take tennis. Before the lob of the ball, the player executes a series of deliberate preparations (e.g., making arrangements with partners and disciplining oneself to practice). From that point on, much of the game is a largely autonomous activity conducted without the planning of each move. As any coach or athlete will attest, the very essence of learning a sport is the intentional acquisition of automaticity, to make the moves seem like second nature.

If we did not economize on our cognitive energy, we would be overwhelmed to the point of near paralysis by everyday demands. Just imagine having to attend to every little bit of life's choreography, such as brushing our teeth, hailing a cab, keeping our temper in check, or obeying the speed limit. Indeed, much of one's talent as a tennis player—and much of one's moral responsibility as a citizen—is about cultivating the right set of "automatic" actions. "Virtues are formed in man by his doing the actions," Aristotle famously proclaimed.[28]

It would therefore be incorrect to think of this kind of freedom as all or nothing, black or white. We should regard it, perhaps, as a mosaic comprising black, white, and gray elements. At times, certain dimensions of our behavior are under conscious control—especially when we have difficult decisions to make, when we plan or when much is at stake—whereas at other times consciousness is bypassed. In the end, it is likely that almost every act emerges from an amalgam of conscious and unconscious processes that assert themselves to varying degrees under the circumstances at the moment. As long as human beings possess conscious mental states that can bring about behavior and self-control, then the law in particular and our moral sense in general need not be radically revised.

HARD determinists such as Greene and Cohen agree that we can use conscious thought to control our actions. Still, they want to purge the law of retributive punishment because they view criminals as "victims of neuronal circumstances" and thus incapable of true choice. Their ideal vision of how the criminal justice system should work—a world without "just desert"—raises a set of interesting questions: Are we even capable of removing blame from the legal domain? Could we pry it loose from our entrenched notions of how human beings operate?

The prospect of a world without moral responsibility collides head-on with our innate sense that people can choose freely. By the age of five, most children perceive others' actions in terms of their intentionality and agency. In one representative experiment, kindergartners observed an investigator sliding open the lid of a box, putting her hand in, and touching the bottom of the box. Asked by the experimenter if she had to touch the bottom or whether she "could have done something else instead," the vast majority of children believed that yes, she could have done something else. But when the experimenter placed a ball on the lid and slid open the lid, and the ball fell to the bottom of the box, only a few of the children said that the ball itself could have "done something else instead."[29]

The intuition that people can do something else persists with age. Adults across cultures, religions, and countries steadfastly reject the idea that the decisions we make are fixed in such a manner that no other actions are possible.[30]

Just as consistently, human beings are highly attuned to notions of fairness. Communities buzz with gossip about who did what scandalous thing to whom; people keep score ("I did two favors for you, but you did only one for me") and punish cheaters who fail to reciprocate. Indigenous people in the Amazonian rain forest are as diligent about detecting free riders and are as ready to punish offenders as are college students in the United States, Europe, and Asia. In his 1991 book *Human Universals*, anthropologist Donald E. Brown presented a comprehensive survey of moral concepts shared across the globe. Among them are proscriptions against murder and rape, as well as redress of wrongs.[31]

In an exhaustive review of anthropological data, psychologists Jonathan Haidt and Craig Joseph found that within all human cultures, individuals tend to react with quick, automatic feelings of anger, contempt, and indignation when they see people causing suffering. The universality of these responses strongly suggests that intuitions about "fairness, harm, and respect for authority has been built into the human mind by evolution," say Haidt and Joseph. "All children who are raised in a reasonable environment will come to develop these ideas, even if they are not taught by adults."[32]

The behavioral economics lab is a showcase for fairness attitudes as well. Daniel Kahneman and colleagues documented how study participants voluntarily incurred a penalty in order to punish others for unfair behavior. It is even more striking that they found evidence that people will pay to punish a third person even when participants are not directly involved in the transaction if they perceive that the third person's unfair behavior was intentional.[33]

The stirrings of moral sentiment begin early on. In a series of studies, psychologists Karen Wynn and Paul Bloom found that toddlers who watched a puppet show in which one puppet "stole" a ball and another puppet returned the ball to its rightful owner were far more likely to give candy to the helper puppet than to the "bad" puppet and preferentially take it away from the "bad" puppet.[34]

More generally, violations of fairness arouse emotions that trigger the urge to retaliate, especially when the violator intentionally injures an innocent person. Social psychologist Philip Tetlock and collaborators presented subjects with a scenario involving a brutal assault that left the victim permanently brain damaged. No matter what happened to the aggressor—in one hypothetical scenario he suffered a painful accident; in another he was cured with medication—subjects' desire for justice was unchanged. Merely rendering him harmless or causing pain accidentally was not sufficient.[35] This is why certain nations direct their law enforcement officials to pursue octogenarian Nazi war criminals living quietly in South America and elsewhere even though they no longer pose a danger. A quickening in our marrow compels us to balance the moral ledger.

Intrinsic to the idea of retribution is that people must suffer in proportion to the suffering they inflicted. Jonathan Haidt and colleagues showed clips from Hollywood films that portrayed injustice (one involving the rape and murder of a child, and another in which a slave's foot is mutilated by a slave catcher). They next gave subjects a variety of endings and asked which one was the most "satisfying." Among the alternative endings was the "revenge" option: The grieving mother violently kills her daughter's rapist; the hobbled slave chops off part of the foot of the man who mutilated him. In the "catharsis" option, the mother undergoes "primal-scream" therapy; the slave chops wood while visualizing the slave catcher's foot. In the "forgiveness" ending, the victims join a support group or become more active in church and learn to forgive the transgression that was committed. The viewers derived far less satisfaction from the scenarios in which the victims come to terms with their tragedies and forgive the transgressor. They wanted the perpetrators to pay. And it was most satisfying when the punishment matched the crime. At the same time, the viewers found gratuitous and less satisfying another ending in which the slave retaliated by murdering his catcher.[36]

This result is consistent with the extensive work of psychologists Kevin M. Carlsmith and John M. Darley. In a series of experiments, they found that when it came to sentencing wrongdoers, subjects were very sensitive to the severity of the offense—say, the theft of $100 to feed a starving child versus using the money to create the world's largest margarita—and largely ignored the likelihood that the person would offend again. Subjects punished people exclusively in proportion to the harm done and not for the harm that might be done in the future. "People want punishment to incapacitate and to deter, but their sense of justice requires sentences proportional to the moral severity of the crime," they concluded. Punishment that violated proportionality offended people's intuitive sense of fairness.[37]

Although hard determinists reject the concept of blameworthiness, they still accept the view that censure can have practical value. As any parent knows, well-calibrated disapproval, along with encouragement, is essential to raising children who are sensitive to the rights of others, behave kindly toward the injured and vulnerable, and reciprocate when they've been helped. No society, whether modern or preliterate, can function and cohere unless its citizens exist within a system of personal accountability that stigmatizes some actions and praises others.[38] Punishment signals to the community who should not be trusted, and its severity reflects the enormity of the transgression. But none of these punishment-related functions requires that anyone be blamed. In utilitarian fashion, punishment could be meted out simply to shape future behavior.

Retribution is a thoroughly different animal. In its pristine, theoretical form, retributive practices are triggered by the simple fact that someone who is mentally competent and uncoerced has committed an offense. The point of punishment is to make perpetrators suffer in proportion to the harm that they have already caused the victim and society. Any incidental benefit to the greater society that comes from reinforcing norms or protecting against future crime is irrelevant.[39] But realistically, when retribution is applied in the real world, it inescapably carries great practical value, too.

For one thing, it strengthens a society's shared norms of moral obligation to one another. One of those norms is that victims should be valued as human beings. Consider the following vignette: A serial rapist, John, attacks Mary, is found guilty, and is sent to prison. A few months later, John is treated with "Castrex," a fictional new anti-rape medication guaranteed to permanently eradicate sexually aggressive urges. Castrex works after just a few doses, and several weeks later, John is freed. He is no longer a danger to anyone. His rehabilitation was a success. But John's light punishment would have woeful repercussions for Mary, her family, and the community at large.

When society fails to condemn aggressors or simply slaps them on the wrist, victims feel unavenged and therefore devalued and dishonored. If perpetrators die before they can be judged or are killed in prison before they can be adequately punished, victims and their families feel enraged. Criminals who do not "pay their debt" can spur victims and their families to contemplate private retaliation and sometimes even undertake it. Contrary to common perception, such feelings do not always arise out of a sense of white-hot rage or bloodlust vigilantism. The motivation to give wrongdoers what they deserve can instead be motivated by grief or by a solemn sense of duty to set things right.[40]

Communities, too, resist what they perceive to be inadequate punishment. Clarence Darrow received "stacks" of letters that were, as he described them, "abusive and brutal to the highest degree," sent by those who wanted his clients hanged. Fast-forward to the 2011 trial of Casey Anthony, the twenty-five-year-old Florida woman who never reported her two-year-old daughter missing. Widely believed to have killed her child or at least to have abetted in her death, Anthony received death threats after she was exonerated for the crime of murder. Likewise, jurists often speak of their moral duty to satisfy the victims and their families. When a U.S. district court judge sentenced disgraced New York City financier Bernard Madoff, who swindled thousands of investors of billions of dollars, to a term of 150 years, he explained to the press that the exceptionally long sentence for an elderly man who would probably die within a decade or so was a symbolic way to help the victims heal.[41]

Restoring the social standing of the victim is another vital function of retribution. When the law imposes inadequate punishment on wrongdoers or allows them to escape penalty altogether, the demoralizing message to the community is clear: The victim is so inconsequential that his or her rights, security, and property can be breached with impunity. This is why proportionate punishment must be carried out in full view of the community in open trials, must be announced in the press, and must be preceded by a clear message of society's moral condemnation. Everyone in the community must be reminded that such mistreatment will not be tolerated. In a series of experiments, legal scholar Kenworthey Bilz asked subjects to evaluate the moral worth of a female victim whose attacker was punished for rape and a similar woman whose rapists pled guilty to a lesser offense. Both women were judged by subjects before their rapists were punished. When the subjects learned that the rapists were convicted for rape, they rated the victim as more "respected," "valued," and "admired" than they were before punishment. When the rapists pled to a lesser nonsexual crime, subjects ranked the victim's social standing lower than in the prepunishment phase.[42]

What happens when the community is prevented from administering justice? In the mid-1960s, sociologist Melvin Lerner developed the "belief in a just world" hypothesis.[43] According to this hypothesis, all of us harbor a strong need to believe that the world is a place where people get what they deserve, a place where actions have predictable consequences. A just-world belief is akin to a "contract" with the world regarding the outcomes of our conduct: We do the right thing, and we are rewarded—or, at least, we know largely what to expect. Lerner understood that the notion of a truly just world is illusory, yet he posited that it helps us to plan our lives and achieve our goals.

In one of his seminal experiments, Lerner asked subjects to observe a ten-minute video of a fellow student as she underwent a learning experiment involving memory. The student was strapped into an apparatus sprouting electrode leads and allegedly received a painful shock whenever she answered a question incorrectly (she was not receiving real shocks, of course, but believably feigned distress as if she were). Next, the researchers split the observers into groups. One was to vote on whether to remove the victim from the apparatus and reward her with money for correct answers. All but one voted to rescue her. The experimenters told another group of observers that the victim would continue to receive painful shocks; there was no option for compensation. When asked to evaluate the victim at this point, subjects in the victim-compensated condition rated her more favorably (e.g., more "attractive," more "admirable") than did subjects in the victim-uncompensated condition, in which the victim's suffering was greater.

Lerner's conclusion is disturbing. "The sight of an innocent person suffering without possibility of reward or compensation motivated people to devalue the attractiveness of the victim in order to bring about a more appropriate fit between her fate and her character," he wrote. In other words, when subjects' intuitions of justice are satisfied, their belief in a just world is supported. But when subjects (read: society) are prevented from restoring justice, they blame the victim. Somehow, the reasoning goes, she must have asked for it.[44]

Beyond the victims, the legal system itself suffers when criminals are not punished, although not just any punishment will do. The penalty must seem proportionate to the offense. If it is perceived as skewed in either direction—too lenient or too harsh—the law can lose its moral force. Recall Rodney King, the man who led California police on a high-speed chase before being caught and beaten severely. Shortly after the 1992 acquittal of the police officers involved in the assault, half of all Californians surveyed said that they had lost confidence in the court system. In a system perceived as unfair, juries may ignore judges' instructions, and police officers may impose their own judgment on whether to arrest, trump up charges, or abuse suspects. For their part, witnesses may rebel by refusing to participate in investigations or to testify. Researchers have observed that subjects are more willing to commit minor offenses such as traffic violations, petty theft, and copyright violations when laws more generally don't conform to their sense of right and wrong. Jurors are more susceptible to nullification—that is, to acquitting defendants who are legally guilty—when the verdict dictated by law is contrary to their sense of justice, morality, or fairness. Examples of perceived unfairness include suspicions of judges' failure to admit evidence, prosecutorial suppression of evidence, police misconduct, or police lying on the stand.[45]

Finally, our intuitions about justice are strong motivators for social change. The victims' rights movement grew out of a powerful sense of unfinished business. Unmet justice alienates juries, judges, and prosecutors. Thus undermining the authority of the law, an institution that depends for its smooth functioning on the good faith of its participants. Pragmatic efforts aimed solely at reducing future crime (and not at all on exacting retribution for past bad deeds) would leave many victims questioning the moral credibility of the law. Those victims don't necessarily seek harsh treatment for their violators though they surely want judges to administer appropriate punishment. What they want most is for the law to acknowledge that what happened to them was wrong and morally offensive.[46] Indeed, the idea of a victim impact statement was developed so that the sentencing judge could hear directly about the anguish that victims and their loved ones had suffered. Some victims want an apology from their violators—not in place of a penalty, but in addition to it.

In summary, some scientists say that they look forward to a day when neuroscience will explain the mechanical workings of the brain so thoroughly that it will be impossible for society to continue to ignore the "fact" that people do not choose their behavior and therefore should be absolved of blame. This vision seems to be on a collision course with the psychological and social meanings that retribution holds for victims, their loved ones, and societies. Victims of crime are exquisitely attuned to society's regard for the people who have violated their rights. Unless authorities mete out deserved punishment, victims feel devalued and suffer a loss of moral standing in their community. Society is also harmed when citizens don't see justice being served and thereby lose faith in the moral authority of the law.

PHILOSOPHERS have wrestled for centuries with the question whether moral responsibility can exist in a world in which our every action is predetermined by a cascade of events leading up to it. Scholars have yet to pin it to the mat, but as legal scholar Stephen Morse reminds us, the law does not require victory.[47] To regard individuals as responsible agents, the law requires that they be able to use conscious thought to control their actions, know what they are doing, and understand the rules. That a long chain of physical causes precedes a crime does not undermine the law's capacity, and duty, to blame and punish.

But should it? Clarence Darrow thought so. "Is Dickey Loeb to blame because out of the infinite forces that conspired to form him . . . he was born without [inner emotion]?" If he is, "then there should be a new definition for justice," Darrow intoned.[48] Many neuroscientists concur and advance a utilitarian model of justice dedicated solely to preventing crime through deterrence, incapacitation, and rehabilitation. What's more, these scientists seem confident that as the general public becomes more familiar with the latest discoveries about the workings of the brain, it will inevitably come to accept their view on moral agency.

This, however, seems too extravagant a hope. The high degree of consensus across cultures regarding the value of fair punishment suggests that human intuitions about fairness and justice are so deeply rooted in evolution, psychology, and culture that new neuroscientific revelations are unlikely to dislodge them easily, if at all. This is not because people are immune to change. On the contrary, attitudes can shift over time, and recent history bears this out. Within the last two centuries alone, we have witnessed profound moral transformations, ranging from the abolition of slavery to legal protections against racial and sexual inequality and to the endorsement of same-sex marriage by millions. Yet these milestones of moral progress would not have come about at all but for the universal human hunger for fairness and justice.

For the sake of argument, though, let's say that officials were to call a moratorium on blame next week. Would the universe eventually become a more humane place for criminals, as the neurodeterminists claim?[49] A kinder domain, perhaps, for us all? Is it conceivable that victims' attitudes could morph so drastically that a few Castrex pills for their rapists would seem like a reasonable way to handle the matter? This is ultimately an empirical question, but we suspect that the abolition of blame would have serious adverse consequences.

For one thing, a blameless world would be a very chilly place, inhospitable to the warming sentiments of forgiveness, redemption, and gratitude. In a milieu where no individuals are accountable for their actions, the so-called moral emotions would be unintelligible.[50] If we no longer brand certain actions as blameworthy and punish transgressors in proportion to their crimes, we forgo precious opportunities to reaffirm the dignity of their victims and to inculcate a shared vision of a just society. By failing to reflect the moral values of the citizenry, which encompass fair punishment, the law would lose some, if not most, of its authority.

The hard determinist vision is locked in an age-old battle with the view that we are morally accountable as long as we are able to reason, save for such exceptions as people with certain forms of brain damage or severe mental illness, and act in accord with our conscious desires. Neuroscience does not hold the key. Instead, we must ask what kind of freedom—ultimate or ordinary—is necessary for moral accountability. The answer will arise out of our intuitions of fairness, not from a lab. What type of neuroscience experiment could even begin to settle it anyway?[51] If there is such a brilliant crucible sketched out in some researcher's notebook, it has yet to be performed. Until then, debates surrounding the value of just deserts must weigh the potential harms that they pose to offenders, the society, and victims against the benefits they afford.

Brain research will continue to yield vast knowledge about the science of thinking and decision making. It will help to explain how we deliberate, weigh options, intend our actions, reflect on our desires, and modify our behaviors on the basis of foreseeable consequences. Brain science will also show us why some people are not able to do this well and, ideally, devise ways to help them. But brain science can never show us that it is unfair or immoral to blame or punish people in a determined world. This means that the contested future of blame will endure as a problem custom-made for the deliberative and conscious creatures that we are.

Epilogue

MIND OVER GRAY MATTER

B RAIN IMAGING, the iconic tool of neuroscience, finds itself at the eye of a perfect storm of seduction. Riding one current is the glamour of a sophisticated and exciting new technology. Borne aloft on another is the brain itself, an organ of great moment and mystery. On a third front floats an overly simplified brain-to-behavior narrative, all rendered in stunning biological portraiture. It is easy to see how nonprofessionals, and an occasional expert, tossed by these powerful swells, can get swept away.

We wrote this book to serve as an anchor. Our project is not a critique of neuroscience or of its signature instrument, brain imaging. It is foremost an exposé of mindless neuroscience: the oversimplification, interpretive license, and premature application of brain science in the legal, commercial, clinical, and philosophical domains.[1] Secondarily but importantly, it is also a critique of the increasingly fashionable assumption that the brain is the most important level of analysis for understanding human behavior, and that the mind—the psychological products of brain activity—is more or less expendable.

We are unreserved champions of neurotechnological progress. We are certain that brain imaging techniques and other exciting developments in neuroscience will further elucidate the relationship between the brain and the mind. We deeply admire the neuroscientists whose inquiries are yielding new discoveries and, perhaps soon, much-needed treatments. In the preceding chapters, however, we have tried to bring a circumspect view to real-world applications of neuroscience and to speculations about where insights gleaned from brain science may take our society. As we've seen, the illuminated brain cannot be trusted to offer an unfiltered view of the mind. Nor is it logical to regard behavior as beyond an individual's control simply because the associated neural mechanisms can be shown to be "in the brain."

Scans alone cannot tell us whether a person is a shameless liar, loyal to a product brand, compelled to use cocaine, or incapable of resisting an urge to kill. In fact, brain-derived data currently add little or nothing to the more ordinary sources of information we rely on to make those determinations; mostly, they are neuroredundant. At worst, neuroscientific information sometimes distort our ability to distinguish good explanations of psychological phenomena from bad ones.

We don't foresee neuroscience prompting a legal revolution. We agree with Stephen Morse that neuroscience will take its place along with other sciences that had their moment in the courtroom: Freudian analysis, behavioral psychology, the Chicago school of sociology, and the promise of genetic explanations. "The only thing different about neuroscience," according to Morse, "is that we have prettier pictures and it appears more scientific."[2] With the probable exception of Freudianism, the other disciplines have indeed made courtroom contributions to understanding why people act as they do. But they have hardly supplanted the bread-and-butter tools of the law, such as witness reports and cross-examination.

Neuroscientists cannot yet forge tight causal links between brain data and behavior. Until they can shed light on the measurable attributes that the law regards as important for culpability—who is and who isn't responsive to reason—the rhetorical value of brain images will greatly outstrip their legal relevance. Within the law, ascriptions of criminal and moral responsibility do not hinge on what caused the bad behavior, but on whether wrongdoers possessed sufficient rational capacity to have been influenced by foreseeable consequences and to alter their behavior accordingly. This is why it has been said that "actions speak louder than images" in today's courtrooms, as well they ought to.[3]

Brain-based explanations for excessive appetites and for social behaviors that elide the crucial psychological, social, and cultural levels of analysis fall into the trap of neurocentrism. Therefore, they are virtually guaranteed to be impoverished explanations. Although scientists can describe human behavior on a number of different levels—the neuronal, the mental, the behavioral, the social—they are not close to bridging the yawning gap between the physical and psychological. The brain enables the mind and thus the person. But neuroscience cannot yet, if ever, fully explain how this happens.

As brain science continues to permeate the culture, neuroliteracy becomes ever more important. Neuroscience is one of the most important intellectual achievements of the past half century, but it is young and still getting its bearings. To demand the wrong things of brain science, to overpromise on what it can deliver, and to apply its technology prematurely will not only tarnish its credibility, it will also risk diverting crucial and limited resources, including federal funding for research, into less profitable ventures and blind alleys.

Skilled science journalists and bloggers, as well as neuroscientists and philosophers who write for the public and neuroethicists (a hybrid sort of scholar with training in both practical philosophy and science), now see part of their jobs as protecting the integrity of neuroscience from the growing legion of brain overclaimers.[4] Responsible translators of neuroscience encourage a healthy skepticism, cautioning judges and policy makers in particular that brain activity elicited under narrow experimental conditions cannot currently yield enough information to explain or predict human behavior in the real world, let alone inform the design of social policy.

Crucial lessons in neuroliteracy must also inculcate the importance of distinguishing the questions that neuroscience is equipped to answer from those that it is not. The job of neuroscience is to elucidate the brain mechanisms associated with mental phenomena, and when technical prowess is applied to the questions it can usefully address, the prospects for conceptual breakthroughs and clinical advances are bountiful. Asking the wrong questions of the brain, however, is at best a dead end and at worst a misappropriation of the mantle of science.

Recall neuroscientist Sam Harris, whom we cited early in this book. "The more we understand ourselves at the level of the brain," he wrote, "the more we will see that there are right and wrong answers to questions of human values."[5] How so? Neuroscience can help answer questions about the neural processes involved in moral decision making, but it is not at all evident how such discoverable facts could ever constitute a prescription for how things *should* be. Surely, empirical facts can help us act more effectively on our values—if we want to rehabilitate prisoners more effectively, data on new therapies are essential. And neuroscience may be able to offer guidance in this regard. But whether we should jettison the practice of retribution on moral grounds is not a question that science, neuroscience included, can answer. Indeed, history is replete with feckless and at times bloody attempts at social engineering through biology. Then and now, it is a serious mistake to think that one can erect an ethical system based on science alone; philosophers call this confusion between "ought" and "is" the naturalistic fallacy.

Nonetheless, the great cultural authority of brain science renders it vulnerable to conscription in the service of one or another political or social agenda. The framing of addiction as a brain disease to attract more funding for research and better services for drug abusers might seem benign; in most cases it is surely well intentioned. But that perspective sorely misrepresents the multilayered nature of addiction and risks distracting clinicians from the most promising kinds of interventions. The same is true to some extent for many other psychological maladies (including psychopathy, the condition that likely afflicted murderer Brian Dugan), which, although surely rooted in brain dysfunction at some level, can be fully understood only by also accommodating the idiom of motives, feelings, thoughts, and decisions.

Likewise, invoking brain science as a rationale for negating blame and abandoning punishment practices is misguided. Neuroscience itself is not a threat to personhood. It will help explain how human agency works, but it will not explain it away. A strictly utilitarian model of justice—one in which we punish people solely because aversive stimuli make society work better, not because blame is truly deserved—has its merits and its shortcomings, depending on your view. But whether human beings who live in a material world can also be moral agents is not a question that brain science can resolve. Not unless, that is, investigators can show something truly spectacular: that people are *not* conscious beings whose actions flow from reasons and who are responsive to reason. True, we do not exert as much conscious control over our actions as we think we do, but this doesn't mean that we are powerless.

In 1996, author Tom Wolfe penned a widely cited essay, "Sorry, but Your Soul Just Died." Neuroscience, he wrote, was on "the threshold of a unified theory that will have an impact as powerful as that of Darwinism a hundred years ago."[6] Almost two decades later, the excitement surrounding neuroscience continues to grow, as well it should. But the promise of a unified theory in the foreseeable future is an illusion. As with sociobiology and the genomic revolution— two valuable conceptual legacies of Darwinism—we should extract the wisdom neuroscience has to offer without asking it to explain all of human nature.

In 2011, science writer David Dobbs recounted a sobering encounter at a gathering of neuroscientists. He asked them, "Of what we need to know to fully understand the brain, what percentage do we know now?"[7] They all responded with figures in the single digits. This humbling estimate will improve with time, of course. Brain imaging will become more precise; new technologies are yet to be unveiled or even envisioned. Yet no matter how dazzling the fruits of inquiry or how clever the means by which they are obtained, it is our values that will guide us in implementing them for good or for bad. The danger lies in muddling those values under the pretense of following where neuroscience supposedly leads us.

To some neuroscientists and philosophers, you may be nothing more than your brain—and of course, without a brain there is no consciousness at all. But to you, you are a "self," and to others you are a person—a person whose brain affords, at once, the capacity for decisions, the ability to study how decisions happen, and the wisdom to weigh the responsibilities and freedoms that these decisions make possible.

ACKNOWLEDGMENTS

We are indebted to many colleagues for their knowledge and insight. Stephen Morse and Hank Greely were invaluable to us in writing chapters on legal matters. Peter Bandettini was the most patient tutor on fMRI that two nonexperts could ever have. Tamler Sommers helped guide us through the rocky shoals of philosophy. Mark Kleiman was a tireless reader of several drafts of the chapter on addiction.

We are fortunate to have superb colleagues who graciously shared their wisdom. They included Craig Bennett, Marc Blitz, Paul Bloom, Nancy Campbell, Christopher Chabris, David Courtwright, Franz Dill, Roger Dooley, Robert DuPont, Steven Erickson, Martha Farah, Nita Farahany, Nathan Greenslit, Stephan Hamann, Wray Herbert, Bryce Huebner, Steven Hyman, Jerome Jaffe, Adam Keiper, Adam Kolber, Annie Lang, Carl Marci, Lori Marino, Richard McNally, Barbara Mellers, Jonathan Moreno, Emily Murphy, Eric Nestler, Joshua Penrod, Steven Pinker, David Pizarro, Russell Poldrack, Anthony Pratkanis, Eric Racine, Richard Redding, Kevin Sabet, Charles Schuster, Roger Scruton, Francis X. Shen, Raymond Tallis, Carol

Tavris, Nehal Vadhann, Edward Vul, Amy Wax, and Christopher E. Wilson.

Special thanks go to Alan I. Leshner who readily made time to talk with an avowed critic of his vision of drug addiction. We are most grateful to Karlyn Bowman, Francis Kissling, Christine Rosen, and Alan Viard for reading draft chapters and for rich discussions about the cultural aspects of neuroscience. Our regular reading of the uniformly excellent online commentaries by *Mind Hacks,* Neuroskeptic, and Neurocritic—neurobloggers all—flagged important trends in neuroscience. We also thank Cheryl Miller and Susan Adams for their valuable editorial work. Wistar Wilson, Catherine Giffin, Elizabeth DeMeo, and Brittany French were our astute research assistants, and Gerry Ohrstrom provided generous support for Sally Satel's research at AEI. Any errors are our own.

This book would not have been written without Lara Heimert, publisher of Basic Books and our editor. We are grateful to her for seeing promise in our project from our first meeting and for her wise counsel. Much appreciation is also due to Charles Eberline, Melody Negron, and Roger Labrie for their excellent editing. Michael Carlisle of Inkwell Management is the most generous and encouraging agent anyone could hope for.

Arthur Brooks, president of the Washington D.C., think tank, the American Enterprise Institute, oversees a wide-ranging intellectual environment that allowed our ideas—neither traditionally political nor discretely policy-oriented—to blossom into a book.

NOTES

Introduction

1. "Research into Brain's 'God Spot' Reveals Areas of Brain Involved in Religious Belief," *Daily Mail*, March 10, 2009, http://www.dailymail.co.uk/sciencetech/article-1160904/Research-brains -God-spot-reveals-areas-brain-involved-religious-belief.html; Susan Brink, "Brains in Love," *Los Angeles Times*, July 30, 2007, http://articles.latimes.com/2007/jul/30/health/he-attraction30; Gabrielle LeBlanc, "This Is Your Brain on Happiness," *Oprah Magazine*, March 2008, http://psyphz .psych.wisc.edu/web/News/OprahMar2008.pdf; Matt Danzico, "Brains of Buddhist Monks Scanned in Meditation Study," BBC, April 23, 2011, http://www.bbc.co.uk/news/world-us-canada -12661646; "Addiction, Bad Habits Can 'Hijack' the Brain," ABC News, January 31, 2012, http:// abcnews.go.com/GMA/MindMoodNews/addictions-hardwired-brain/story?id=9699738; Alice Park, "The Brain: Marketing to Your Mind," *Time*, January 29, 2007, http://www.time.com/time /magazine/article/0,9171,1580370-1,00.html; Chris Arnold, "Madoff's Alleged Ponzi Scheme Scams Smart Money," NPR, December 6, 2008, http://www.npr.org/templates/story/story.php?sto ryId=98321037 (Bernie Madoff financial fiasco); Sharon Begley, "Money Brain: The New Science Behind Your Spending Addiction," *Newsweek*, October 30, 2011, 2012, http://www.thedailybeast .com/newsweek/2011/10/30/the-new-science-behind-your-spending-addiction.html (U.S. debt limit nail-biter of 2011); "You Love Your iPhone. Literally," *New York Times*, September 30, 2011, http://www.nytimes.com/2011/10/01/opinion/you-love-your-iphone-literally.html (love of iPhones); Christine Morgan, "Addicted to Thrills: Why We Love Scary Movies," *Daily Mail*, July 3, 2011, http://www.mydaily.co.uk/2011/03/07/why-some-people-are-thrill-seekers/ (affinity for horror movies); David J. Linden, "Anthony Weiner, Straus-Kahn, Arnold Schwarzenegger: Are They Just Bad Boy Politicians or Is It Their DNA?," *Huffington Post,* June 14, 2011, http://www .huffingtonpost.com/david-j-linden/notorious-politicans_b_876428.html#s291507& title=Anthony_Weiner (sexual indiscretions of politicians); Chris Mooney, *The Republican Brain: The Science of Why They Deny Science—and Reality* (Hoboken, NJ: Wiley, 2012) (conservatives' dismissal of global warming); and C. R. Harrington et al., "Activation of the Mesostriatal Reward Pathway with Exposure to Ultraviolet Radiation (UVR) vs. Sham UVR in Frequent Tanners: A Pilot Study," *Addiction Biology* 3 (2011): 680–686.

2. For documentation of the status and growth of neuroeconomics, see also Josh Fischman, "The Marketplace in Your Brain," *The Chronicle Review,* September 24, 2012, http://chronicle .com/article/The-Marketplace-in-Your-Brain/134524/; "What Is Neurohistory," Neurohistory, http://www.neurohistory.ucla.edu/neurohistory-web-about. Notably, neuromusicology has produced some interesting theoretical work. For a rich perspective on neuroscience and music, see Daniel Levitin, *This Is Your Brain on Music: The Science of a Human Obsession* (New York: Dutton, 2006). For a virtuoso synthesis, see Eric R. Kandel, *The Age of Insight: The Quest to Understand the Unconscious in Art, Mind, and Brain, from Vienna 1900 to the Present* (New York: Random House, 2012). See also Uri Hasson et al., "Neurocinematics: The Neuroscience of Film," *Projections* 2, no. 1 (2008): 1–26, http://www.cns.nyu.edu/~nava/MyPubs/Hasson-etal_Neuro Cinematics2008.pdf. Paul M. Matthews and Jeffrey McQuain, *The Bard on the Brain: Understanding the Mind Through the Art of Shakespeare and the Science of Brain Imaging* (New York: Dana Press, 2003); Patricia Cohen, "Next Big Thing in English: Knowing They Know That You Know," *New York Times,* March 31, 2010, http://www.nytimes.com/2010/04/01/books/01lit.html ?pagewanted=all; and Paul Harris and Alison Flood, "Literary Critics Scan the Brain to Find Out Why We Love to Read," *Guardian,* April 11, 2010, http://www.guardian.co.uk/science/2010/apr /11/brain-scans-probe-books-imagination. For a discussion of competing views on the value of "neuro-lit," see Roger Scruton, "Brain Drain: Neuroscience Wants to Be the Answer to Everything—It Isn't," *Spectator,* March 17, 2012, http://www.spectator.co.uk/essays/all/7714533/brain -drain.thtml; "Can Neuro-Lit Save the Humanities?," Room for Debate, *New York Times,* April 5, 2012, http://roomfordebate.blogs.nytimes.com/2010/04/05/can-neuro-lit-crit-save-the -humanities/; and Alva Noë, "Art and the Limit of Neuroscience," Opinionator, *New York Times,* December 4, 2011, http://opinionator.blogs.nytimes.com/2011/12/04/art-and-the-limits -of-neuroscience/?pagemode=print.

3. On the brain as a cultural artifact, see, for example, Olivia Solon, "3D-Printed Brain Scan Just One Exhibit at London 'Bio-Art' Show," *Wired,* July 20, 2011, http://www.wired.co.uk/news/ar chive/2011-07/20/art-science-gv-gallery; Bill Harbaugh, "Bachy's Figured Maple Brains," personal website, http://harbaugh.uoregon.edu/Brain/Bachy/index.htm; and Sara Asnagi, "What Have You Got in Your Head? Human Brains Made with Different Foods," Behance, August 1, 2012, http:// www.behance.net/gallery/What-have-you-got-in-your-head/614949. For a cultural analysis of a form of brain scan called positron emission tomography, see Joseph Dumit, *Picturing Personhood: Brain Scans and Biomedical Identity* (Princeton, NJ: Princeton University Press, 2003). See also Tom Wolfe, *I Am Charlotte Simmons* (New York: Farrar, Straus and Giroux, 2004), 392; Ian McEwan, *Saturday* (New York: Nan A. Talese, 2005); and A. S. Byatt, "Observe the Neurones: Between, Below, Above John Donne," *Times Literary Supplement,* September 22, 2006. The quotation about Andy Warhol is from Jonah Lehrer, "The Rhetoric of Neuroscience," *Wired,* August 11, 2011, http://www.wired.com/wiredscience/2011/08/the-rhetoric-of-neuroscience/. Lehrer was fired from the *New Yorker* for fabricating quotes by songwriter Bob Dylan, but to our knowledge, this quotation is entirely original to him.

4. A number of excellent commentaries on the overselling of "neurohype" have been published within the last few years. See Diane M. Beck, "The Appeal of the Brain in the Popular Press," *Perspectives on Psychological Science* 5 (2010): 762–766. Eric Racine et al., "Contemporary Neuroscience in the Media," *Social Science and Medicine* 71, no. 4 (2010): 725–733; Julie M. Robillard and Judy Illes, "Lost in Translation: Neuroscience and the Public," *Nature Reviews Neuroscience* 12 (2011): 118; Matthew B. Crawford, "On the Limits of Neuro-Talk," *The New Atlantis,* no. 19, Winter 2008, http://www.thenewatlantis.com/publications/the-limits-of-neuro-talk; Raymond, *Aping Mankind: Neuromania, Darwinitis, and the Misrepresentation of Humanity* (Durham, UK: Acumen, 2011); Alva Noe, *Out of Our Heads: Why You Are Not Your Brain and Other Lessons from the Biology of Consciousness* (New York: Hill and Wang, 2010); Paolo Legrenzi and Carlo Umilta, *Neuromania: On the Limits of Brain Science* (Oxford: Oxford University Press, 2011);

and Gary Marcus, "Neuroscience Fiction," *New Yorker*, December 2, 2012, http://www.newyorker.com/online/blogs/newsdesk/2012/12/what-neuroscience-really-teaches-us-and-what-it-doesnt.html. On brain images as the symbol of science, see Martha J. Farah, "A Picture Is Worth a Thousand Dollars," *Journal of Cognitive Neuroscience* 21, no. 4 (2009): 623–624. Popular interest in the brain predated fMRI. In the 1980s, an earlier form of brain imaging called positron emission tomography (PET) also produced dazzling brain pictures, but expense and radioactivity limit its use.

5. Tom Wolfe, "Sorry, but Your Soul Just Died," in *Hooking Up* (New York: Picador, 2000), 90. See also Wolfe's commencement speech expressing the cultural significance of neuroscience in Jacques Steinberg, "Commencement Speeches," *New York Times*, June 2, 2002, http://www.nytimes.com/2002/06/02/nyregion/commencement-speeches-along-with-best-wishes-9-11-is-a-familiar-graduation-theme.html?pagewanted=all&src=pm; and Zack Lynch, *The Neuro Revolution: How Brain Science Is Changing Our World* (New York: St. Martin's Press, 2009).

6. Roberto Lent et al., "How Many Neurons Do You Have? Some Dogmas of Quantitative Neuroscience Under Revision," *European Journal of Neuroscience* 35, no. 1 (2012): 1–9.

7. Philosophers call the misplaced faith in the trustworthiness of our perceptions "naive realism." On Galileo, see Gerald James Holton, *Thematic Origins of Scientific Thought: Kepler to Einstein* (Cambridge, MA: Harvard University Press, 1988), 43–44.

8. Eric Racine has coined the term "neuro-realism" to denote that what we see in a brain scan makes it somehow more real or true than if a visual image were not invoked. See Eric Racine, Ofek Bar-Ilan, Judy Illes, "fMRI in the Public Eye," *Nature Reviews Neuroscience* 6, no. 2 (2005): 159–164, 160.

9. Marco Iacoboni et al., "This Is Your Brain on Politics," *New York Times*, November 11, 2007, http://www.nytimes.com/2007/11/11/opinion/11freedman.html?pagewanted=all.

10. Semir Zeki and John Paul Romaya, "Neural Correlates of Hate," *PLoS One* 3, no. 10 (2008). These neuroscientists ostensibly revealed the "hate circuit," as a press release from University College put it; article available at http://www.plosone.org/article/info%3Adoi%2F10.1371%2Fjournal.pone.0003556; Graham Tibbetts and Sarah Brealey, "'Hate Circuit' Found in Brain," *The Telegraph*, October 28, 2008, http://www.telegraph.co.uk/news/newstopics/howaboutthat/3274018/Hate-circuit-found-in-brain.html. David Robson, "'Hate' Circuit Discovered in Brain," *New Scientist*, October 28, 2008, http://www.newscientist.com/article/dn15060-hate-circuit-discovered-in-brain.html.

11. Andreas Bartels and Semir Zeki, "The Neural Basis of Romantic Love," *NeuroReport* 11, no. 17 (2000): 3829–3834; William Harbaugh, Ulrich Mayr, and Dan Burghart, "Neural Responses to Taxation and Voluntary Giving Reveal Motives for Charitable Donations," *Science* 316, no. 5831 (2007): 1622–1625.

12. For a good overview of brain overclaiming written by neuroscientists, see Garret O'Connell et al., "The Brain, the Science and the Media: The Legal, Corporate, Social and Security Implications of Neuroimaging and the Impact of Media Coverage," *European Molecular Biology Organization Reports* 12 no. 7 (2011): 630–636. Neuroscience blogger Vaughn Bell notes, "To fully understand what happens during a brain imaging experiment you need to be able to grasp quantum physics at one end, to philosophy of mind at the other, while travelling through a sea of statistics, neurophysiology and psychology. Needless to say, very few, if any scientists can do this on their own. . . . Under the sheer weight of conceptual strain, journalists panic, and do this: 'Brain's adventure centre located.'" Vaughn Bell, "The fMRI Smackdown Cometh," Mind Hacks, June 26, 2008, http://mindhacks.com/2008/06/26/the-fmri-smackdown-cometh/. See also Beck, "Appeal of the Brain in the Popular Press." On the labels applied to pop neuroscience, see Raymond Tallis, *Aping Mankind: Neuromania, Darwinitis and the Misrepresentation of Humanity* (Durham, UK: Acumen, 2011); Andrew Linklater, "Incognito: The Secret Lives of the Brain by David Eagleman—Review," *Guardian*, April 23, 2011, http://www.guardian.co.uk/books/2011/apr/24/incognito-secret-brain-david-eagleman ("neurohubris"); and Vaughn Bell, "Don't Believe the Neurohype," May

22, 2008, http://mindhacks.com/2008/05/22/dont-believe-the-neurohype/; and Steven Poole, "Your Brain on Pseudoscience: The Rise of Popular Neurobollocks," *New Statesman*, September 6, 2012, http://www.newstatesman.com/culture/books/2012/09/your-brain-pseudoscience. For an example of dumbing down, see Elizabeth Landau, CNN Health, February 19, 2009, http://articles.cnn.com/2009-02-19/health/women.bikinis.objects_1_bikini-strip-clubs-sexism?_s=PM:HEALTH.

13. Poole, "Your Brain on Pseudoscience."

14. Srinivasan S. Pillay, *Your Brain and Business: The Neuroscience of Great Leaders* (Upper Saddle River, NJ: FT Press, 2011), 15. For examples of brain-based education techniques, see "What Is Brain-Based Learning?," Jensen Learning: Practical Teaching with the Brain in Mind, http://www.jensenlearning.com/what-is-brain-based-research.php; E. E. Boyd, "Why Brain Gyms Might Be the Next Big Business," *Fast Company*, June 16, 2011, http://www.fastcompany.com/1760312/why-brain-gyms-may-be-next-big-business; and Daniel A. Hughes et al., *Brain-Based Parenting: The Neuroscience of Caregiving for Healthy Attachment* (New York: W. W. Norton, 2012). For critiques of "brain-based" education, see Daniel T. Willingham, "Three Problems in the Marriage of Neuroscience and Education," *Cortex* 45 (2009): 544–545; and Larry Cuban, "Brain-Based Education—Run from It," *Washington Post*, February 28, 2011, http://voices.washingtonpost.com/answer-sheet/guest-bloggers/brain-based-education-run-from.html. The quip is from Keith R. Laws, Twitter post, January 28, 2012, 3:13 a.m., http://twitter.com/Keith_Laws/statuses/163218019449962496.

15. David Eagleman, "The Brain on Trial," *The Atlantic*, July/August, 2011, http://www.theatlantic.com/magazine/archive/2011/07/the-brain-on-trial/308520/.

16. David Eagleman, *Incognito: The Secret Lives of the Brain* (New York: Pantheon, 2011), 176.

17. Francis Bacon describes science as a method "which shall analyse experience and take it to pieces." Francis Bacon, *The Plan of the Instauratio Magna*, Bartleby.com, http://www.bartleby.com/39/21.html. William James notes, "A science of the mind must reduce . . . complexities (of behavior) to their elements. A science of the brain must point out the functions of its elements. A science of the relations of mind and brain must show how the elementary ingredients of the former correspond to the elementary functions of the latter." William James, *The Principles of Psychology* (Mineola, NY: Dover, 1950), 28. On the hierarchical levels, see Kenneth S. Kendler, "Toward a Philosophical Structure for Psychiatry," *American Journal of Psychiatry* 162, no. 3 (2005): 433–440; and Carl F. Craver, *Explaining the Brain* (Oxford: Oxford University Press, 2009): 107–162.

18. This analogy is borrowed from David Watson, Lee Anna Clark, Allan R. Harkness, "Structures of Personality and Their Relevance to Psychopathology," *Journal of Abnormal Psychology*, 103 (1994): 18–31.

19. This famously recalcitrant enigma has no solution in sight. Colin McGinn, "Can We Solve the Mind-Body Problem?," *Mind* 98 (1989): 349–366.

20. Sam Harris, *The Moral Landscape: How Science Can Determine Human Values* (New York: Free Press, 2010); Semir Zeki and Oliver Goodenough, *Law and the Brain* (Oxford: Oxford University Press, 2006), xiv; Michael S. Gazzaniga, *The Ethical Brain* (New York: Dana Press, 2005), xv, xix. See also Arne Rasmusson, "Neuroethics as a Brain-Based Philosophy of Life—The Case of Michael S. Gazzaniga," *Neuroethics* 2 (2009): 3–11.

21. Ron Rosenbaum, "The End of Evil? Neuroscientists Suggest There Is No Such Thing. Are They Right?," *Slate*, September 30, 2011, http://www.slate.com/articles/health_and_science/the_spectator/2011/09/does_evil_exist_neuroscientists_say_no_.html.

22. Neuroskeptic, "fMRI Reveals the True Nature of Hatred," October 30, 2008, http://neuroskeptic.blogspot.com/2008/10/fmri-reveals-true-nature-of-hatred.html. The imaging technique was first announced in a 1991 study published in *Science* that demonstrated how a standard MRI scanner could be used to track where oxygenated and deoxygenated blood flowed in the brain. J. W. Belliveau et al., "Functional Mapping of the Human Visual Cortex by Magnetic Resonance Imag-

ing," *Science* 254 (1991): 716–719. For general overview of accomplishments, see Michael Gazzaniga, *The Cognitive Neurosciences*, 4th ed. (Cambridge, MA: MIT Press, 2009). For a de facto requirement in imaging expertise, see Gregory A. Miller, "Mistreating Psychology in the Decades of the Brain," *Perspectives on Psychological Science* 5, no. 6 (2010): 716–743. Harvard psychologist Jerome Kagan notes that graduate students routinely go out of their way to design dissertation topics to include an imaging component: "Neuroscience was High Church and studying the brain was a required ritual for anyone who wished to be ordained into a holy order." Jerome Kagan, *An Argument for Mind* (New Haven, CT: Yale University Press, 2006), 17–18. See also Paul Bloom, "Seduced by the Flickering Lights," *Seed*, June 26, 2006, http://seedmagazine.com/content/article /seduced_by_the_flickering_lights_of_the_brain/.

23. Biologist Steven Rose may have been the first to use this term in the context of cultural preoccupation with the brain. Steven P. R. Rose, "Human Agency in the Neurocentric Age," *EMBO Reports* 6 (2006): 1001–1005. We use the term somewhat differently.

24. David Linden, "'Compass of Pleasure': Why Some Things Feel So Good," NPR, June 23, 2011, http://www.npr.org/2011/06/23/137348338/compass-of-pleasure-why-some-things-feel-so-good.

25. "The urge for retribution depends upon our not seeing the underlying causes of human behavior." Sam Harris, *Free Will* (New York: Free Press, 2012), 55. On the relationship between brain-based explanations and responsibility, see Stephen Morse, "Brain Overclaim Syndrome and Criminal Responsibility," *Ohio State Journal of Criminal Law* 3 (2006): 397–412.

26. Robert M. Sapolsky, quoted in personal comment to Michael Gazzaniga and cited in Michael S. Gazzaniga, *Who's In Charge? Free Will and the Science of the Brain* (New York: Ecco, 2011), 188. See also Sapolsky, "The Frontal Cortex and the Criminal Justice System," *Philosophical Transactions of the Royal Society of London* 359 (2004): 1787–1796.

27. For neuromarketing, see Neurofocus, accessed July 7, 2011, neurofocus.com. For lie-detection services, see No Lie MRI, accessed September 3, 2012, http://www.noliemri.com/. For political consulting, see Westen Strategies, accessed September 3, 2012, http://www.westenstrategies.com/, where it is stated that "to move people, you have to understand the neural networks that connect ideas, images, and emotions in their minds."

Chapter 1

1. Jeffrey Goldberg, "Re-thinking Jeffrey Goldberg," *Atlantic*, July–August 2008, http://www .theatlantic.com/doc/200807/mri/2; http://www.theatlantic.com/daily-dish/archive/2008/06/jeffrey -goldberg-closet-shiite/215362/.

2. Computer programs correct for head movement, swallowing, jaw clenching, breathing, and even pulsation of the carotid artery. The scanner is also claustrophobic. "As many as 20 percent of subjects are similarly affected. Because not everyone can remain relatively relaxed while squeezed inside the tube, fMRI studies are afflicted with a selection bias; the subject sample cannot be completely random, so it cannot be said to represent all brains fairly"; Michael Shermer, "Five Ways Brain Scans Mislead Us," *Scientific American*, November 5, 2008, http:// www.scientificamerican.com/article.cfm?id=five-ways-brain-scans-mislead-us. There are many sounds associated with scanning. The echo planar imaging (EPI; the sequence that does the functional scanning) is like a high-pitched pinging. The metal cleat part is what scanners typically sound like when they are collecting the average anatomical, or structural, template—the magnetic resonance image of the subject's brain at the beginning of the session. A good resource on many MRI sounds is York Neuroimaging Center, "MRI Sounds," https://www.ynic.york.ac.uk /information/mri/sounds/.

3. Goldberg, "Re-thinking Jeffrey Goldberg"; William R. Uttal, *The New Phrenology: The Limits of Localizing Cognitive Processes in the Brain* (Cambridge, MA: MIT Press, 2001); Greg

Miller, "Growing Pains for fMRI," *Science* 320, no. 5882 (2008): 1412–1414, www.scribd.com /doc/3634406/Growing-pains-for-fMRI; Hanna Damasio, "Beware the Neo Phrenologist: Modern Brain Imaging Needs to Avoid the Mistakes of Its Predecessor," *USC Trojan Magazine*, Summer 2006, http://www.usc.edu/dept/pubrel/trojan_family/summer06/BewareNeo.html.

4. For the history of imaging, see Bettyann H. Kevles, *Naked to the Bone: Medical Imaging in the Twentieth Century* (New Brunswick, NJ: Rutgers University Press, 1997). In 1896 a London company marketed anti-X-ray underwear to reassure uneasy Victorian citizens concerned that the new technology would expose sensitive regions of their anatomy to public view; see Brian Lentle and John Aldrich, "Radiological Sciences, Past and Present," *Lancet* 350, no. 9073 (1997): 280–285, http://www.umdnj.edu/idsweb/shared/radiology_past_present.html. On Baraduc, see Elmar Schenkel and Stefan Welz, eds., *Magical Objects: Things and Beyond* (Berlin: Galda and Wilch Verlag, 2007), 140.

5. For an excellent overview of the history of neuroimaging, see *Human Functional Brain Imaging, 1990–2009* (London: Wellcome Trust, 2011), http://www.wellcome.ac.uk/stellent/groups /corporatesite/@policy_communications/documents/web_document/WTVM052606.pdf. Computer-assisted tomography, or computeraxial tomography, is an advanced X-ray technique. In CAT scanners, the technique is used to obtain cross-sectional images of the body. Information provided by an X-ray is processed by a computer to produce a three-dimensional image of the inside of an object from a large series of two-dimensional X-ray images taken around a single axis of rotation.

6. Single-photon emission computed tomography (SPECT) is another technique using radioactive material. See *Human Functional Brain Imaging, 1990–2009* for a complete overview of all functional imaging techniques. PET requires a hugely expensive infrastructure, including a cyclotron to make the tracers. Also, the temporal resolution is very slow; one image is acquired about every thirty seconds. In fMRI, slices are usually obtained every two to four seconds. Rapid decay of the tracer means that experimental sessions must be conducted under time constraints. It takes almost one minute for PET to measure the blood flow to an area, whereas fMRI executes a measurement every two seconds. Also, PET produces blurrier images because of its weaker spatial resolution of six to nine millimeters, compared with three millimeters or fewer for fMRI.

7. The following are excellent general sources on the construction of brain scans: Russell A. Poldrack, Jeanette A. Mumford, and Thomas E. Nichols, *Handbook of Functional MRI Data Analysis* (Cambridge: Cambridge University Press, 2011); Peter Bandettini, ed., "20 Years of fMRI," *Neuroimage* 62, no. 2 (2012): 575–588; and Nikos K. Logothetis, "What We Can Do and What We Cannot Do with fMRI," *Nature* 453 (2008): 869–878. Subjects must remove watches and rings in deference to the magnet, which is about 60,000 times as powerful as the earth's magnetic field. The magnet can wipe credit cards clean, cause unsecured IV poles to rush toward the machine, or dislodge implanted medical devices containing metal, causing them to malfunction and possibly injure surrounding tissues. Robert S. Porter, ed., "Magnetic Resonance Imaging," in *Merck Manual Home Health Handbook*, 2008, http://www.merckmanuals.com/home/special_subjects/common _imaging_tests/magnetic_resonance_imaging.html.

8. Hippocrates, "The Sacred Disease," written ca. 400 BCE, cited in Bob Kentridge, "S2 Psychopathology: Lecture 1," 1995, http://www.dur.ac.uk/robert.kentridge/ppath1.html. On the Epicureans, see generally "Epicurus," *Stanford Encyclopedia of Philosophy*, February 18, 2009, http:// plato.stanford.edu/entries/epicurus/#3. See also Carl Zimmer, *Soul Made Flesh: The Discovery of the Brain and How It Changed the World* (New York: Free Press, 2004).

9. Stanley Finger, *Origins of Neuroscience: A History of Explorations into Brain Function* (Oxford: Oxford University Press, 2001); Raymond E. Fancher, *Pioneers of Psychology,* 3rd ed. (New York: Norton, 1996), 25–26; and Zimmer, *Soul Made Flesh,* 31–41.

10. William James, *Psychology: The Briefer Course* (1892; Mineola, NY: Dover, 2001), 335. Sigmund Freud, too, wanted to represent psychic processes as quantitatively determinate states of

specifiable material particles. Faced with overwhelming technical obstacles, he abandoned neuro-science and turned to the abstract realm of the unconscious. "The intention of this project is to furnish us with a psychology which shall be a natural science: its aim, that is, is to represent psychical processes as quantitatively determined states of specifiable material particles," he wrote in "Project for a Scientific Psychology" (1895), in *The Complete Psychological Works of Sigmund Freud*, trans. James Strachey (London: Hogarth Press, 1886–1899), 1:299.

11. Paul Bloom, "Seduced by the Flickering Lights of the Brain," *Seed,* June 27, 2006, http://seedmagazine.com/content/article/seduced_by_the_flickering_lights_of_the_brain/.

12. Finger, *Origins of Neuroscience,* 32–43.

13. Malcolm MacMillan, *An Odd Kind of Fame: Stories of Phineas Gage* (Cambridge, MA: MIT Press, 2000). Within a small circle of devoted experts, there is controversy over how fully Gage recovered before his death twelve years later and even how severe his postinjury symptoms actually were.

14. See generally John Van Wyhe, *Phrenology and the Origins of Victorian Scientific Naturalism* (Aldershot, UK: Ashgate, 2004). As a footnote to history, the tamping iron (a rod used to compress an explosive charge inside a crevice or hole used in blasting rock) that injured Gage exited "in the neighborhood of Benevolence and the front part of Veneration"; MacMillan, *Odd Kind of Fame,* 350.

15. Max Neuburger, "Briefe Galls an Andreas und Nannette Streicher," *Archiv für Geschichte der Medizin* 10 (1917): 3–70, 10, cited in John Van Wyhe, "The Authority of Human Nature: The Schädellehre [skull reading] of Franz Joseph Gall," *British Journal for the History of Science* 35 (2002): 17–42, 27; Steven Shapin, "The Politics of Observation: Cerebral Anatomy and Social Interests in the Edinburgh Phrenology Disputes," in *On the Margins of Science: The Social Construction of Rejected Knowledge*, ed. R. Wallis (Keele, UK: University Press of Keele, 1979), 139–178. See generally John D. Davies, *Phrenology, Fad and Science* (New Haven, CT: Yale University Press, 1955).

16. Mark Twain, *The Autobiography of Mark Twain*, ed. Charles Neider (New York: HarperCollins, 2000), "startled," 85; "cavity" and "humiliated," 86; "Mount Everest," 87; Delano José Lopez, "Snaring the Fowler: Mark Twain Debunks Phrenology," *Skeptical Inquirer* 26, no. 1 (2002), http://www.csicop.org/si/show/snaring_the_fowler_mark_twain_debunks_phrenology/.

17. Shaheen E. Lakhan and Enoch Callaway, "Deep Brain Stimulation for Obsessive-Compulsive Disorder and Treatment-Resistant Depression: Systematic Review," *BMC Research Notes* 3 (2010): 60, http://www.biomedcentral.com/1756-0500/3/60/. On the possible value of fMRI in predicting recurrence in people who have been treated for depression, see Norman A. S. Farb et al., "Mood-Linked Responses in Medial Prefrontal Cortex Predict Relapse in Patients with Recurrent Unipolar Depression," *Biological Psychiatry* 70, no. 4 (2011): 366–372; and Oliver Doehrmann et al., "Predicting Treatment Response in Social Anxiety Disorder from Functional Magnetic Resonance Imaging," *Archives of General Psychiatry* 70, no. 1 (2013): 87–97. On the use of fMRI in treatment of comatose patients, see David Cyranowski, "Neuroscience: The Mind Reader," *Nature* 486 (2012): 178–180; and Joseph J. Fins, "Brain Injury: The Vegetative and Minimally Conscious States," in *From Birth to Death and Bench to Clinic: The Hastings Center Bioethics Briefing Book for Journalists, Policymakers, and Campaigns*, ed. Mary Crowley (Garrison, NY: Hastings Center, 2008), 15–20, http://www.thehastingscenter.org/Publications/BriefingBook/Detail.aspx?id=2166.

18. Aaron J. Newman et al., "Dissociating Neural Subsystems for Grammar by Contrasting Word Order and Inflection," *Proceedings of the National Academy of Sciences* 107, no. 16 (2010): 7539–7544; Daniel A. Abrams et al., "Multivariate Activation and Connectivity Patterns Discriminate Speech Intelligibility in Wernicke's, Broca's, and Geschwind's Areas," *Cerebral Cortex*, 2012, http://cercor.oxfordjournals.org/content/early/2012/06/12/cercor.bhs165.abstract; Nancy Kanwisher, "Functional Specificity in the Human Brain: A Window into the Functional Architecture of the Mind," *Proceedings of the National Academy of Sciences* 107, no. 25 (2010): 11163–11170;

Lofti B. Merabet and Alvaro Pascual-Leone, "Neural Reorganization Following Sensory Loss—The Opportunity for Change," *Nature Reviews Neuroscience* 11 (2012): 44–53; Luke A. Henderson et al., "Functional Reorganization of the Brain in Humans Following Spinal Cord Injury: Evidence for Underlying Changes in Cortical Anatomy," *Journal of Neuroscience* 31, no. 7 (2011): 2630–2637; M. Ptito et al., "TMS of the Occipital Cortex Induces Tactile Sensations in the Fingers of Braille Readers," *Experimental Brain Research* 184 (2008): 193–200, http://www.ncbi.nlm.nih.gov/pubmed/17717652.

19. Russell Poldrack and Marco Iacoboni debate the scientific legitimacy of the FKF results from the swing-voter study in Adam Kolber, "Poldrack Replies to Iacoboni Neuropolitics Discussion," *Neuroethics & Law Blog,* June 3, 2008, http://kolber.typepad.com/ethics_law_blog/2008/06/poldrack-replie.html, and Adam Kolber, "Iacoboni Responds to Neuropolitics Criticism," *Neuroethics & Law Blog,* June 3, 2008, http://kolber.typepad.com/ethics_law_blog/2008/06/iacoboni-respon.html. On the functions of the amygdala, see Shermer, "Five Ways Brain Scans Mislead Us"; Elizabeth A. Phelps and Joseph E. LeDoux, "Contributions of the Amygdala to Emotion Processing: From Animal Models to Human Behavior," *Neuron* 48, no. 2 (2005): 175–187; and Turhan Canli and John D. E. Gabrieli, "Imaging Gender Differences in Sexual Arousal," *Nature Neuroscience* 7, no. 4 (2004): 325–326.

20. The amygdala contributes to the operation of many tasks because of its involvement with attention, vigilance, and memory. William A. Cunningham and Tobias Brosch, "Motivational Salience: Amygdala Tuning from Traits, Needs, Values, and Goals," *Current Directions in Psychological Science* 21 (2012): 54–59. On amygdala response to pictures of food, see A. Mohanty et al., "The Spatial Attention Network Interacts with Limbic and Monoaminergic Systems to Modulate Motivation-Induced Attention Shifts," *Cerebral Cortex* 18, no. 11 (2008): 2604–2613.

21. Russell Poldrack, "Can Cognitive Processes Be Inferred from Neuroimaging Data?," *Trends in Cognitive Sciences* 10, no. 2 (2006): 59–63.

22. Diane M. Beck, "The Appeal of the Brain in the Popular Press," *Perspectives on Psychological Science* 5 (2010): 762–766; Eric Racine et al., "Contemporary Neuroscience in the Media," *Social Science and Medicine* 71, no. 4 (2010): 725–733; Julie M. Robillard and Judy Illes, "Lost in Translation: Neuroscience and the Public," *Nature Reviews Neuroscience* 12 (2011): 118. On the roles of the insula, see A. D. Craig, "How Do You Feel Now? The Anterior Insula and Human Awareness," *Nature Reviews Neuroscience* 10, no. 1 (2009): 59–70. As a general rule, regions originally believed to implement some unified process will turn out to become more heterogeneous as techniques with better spatial and temporal resolutions arise.

23. Adam Aron et al., "Politics and the Brain," *New York Times,* November 14, 2007, and "Editorial: Mind Games: How Not to Mix Politics and Science," *Nature* 450 (2007), http://www.nature.com/nature/journal/v450/n7169/full/450457a.html; Vaughan Bell, "Election Brain Scan Nonsense," *Mind Hacks* (blog), November 13, 2007, http://www.mindhacks.com/blog/2007/11/election_brain_scan_.html. Neuropundits: Daniel Engber, "Neuropundits Gone Wild," *Slate,* November 14, 2007, www.slate.com/articles/health_and . . . /neuropundits_gone_wild.html. "The scattered spots of activation in a brain image can be like tea leaves in the bottom of a cup, ambiguous and accommodating of a large number of possibilities," says cognitive neuroscientist Martha Farah of the University of Pennsylvania; Adam Kolber, "This Is Your Brain on Politics (Farah Guest Post)," *Neuroethics & Law Blog,* November 12, 2007, http://kolber.typepad.com/ethics_law_blog/2007/11/this-is-your-br.html. For an overview of how to conduct experiments and how (not) to interpret them, see Teneille Brown and Emily Murphy, "Through a Scanner Darkly: Functional Neuroimaging as Evidence of a Criminal Defendant's Past Mental States," *Stanford Law Review* 62, no. 4 (2010): 1119–1208, 1142, http://legalworkshop.org/wp-content/uploads/2010/04/Brown-Murphy.pdf.

24. Rene Weber, Ute Ritterfeld, and Klaus Mathiak, "Does Playing Violent Video Games Induce Aggression? Empirical Evidence of a Functional Magnetic Resonance Imaging Study," *Media Psy-

chology 8 (2006): 39–60. The authors suggest alternative noncausal explanations and also note the caveat that they recruited a nonrepresentative sample. They used advertisements in video-gaming and computer stores. On average, the subjects spent fifteen hours a week gaming. In the press release, Weber is quoted as saying, "Violent video games frequently have been criticized for enhancing aggressive reactions such as aggressive cognitions, aggressive affects or aggressive behavior. On a neurobiological level we have shown the link exists." The key question, not addressed in this small sample study of thirteen men, is whether they actually behaved more aggressively. And again, even if they did, we could not infer, without further research, that the videos were the cause.

25. Another complementary tool is transcranial magnetic stimulation (TMS). Researchers move a wandlike TMS device around the head to produce a painless fluctuating magnetic field that temporarily and reversibly weakens the electric currents in a particular brain region. By effectively removing a region of the brain for a moment, researchers can begin to understand causality by parsing seemingly unitary operations, such as memory, visual perception, and attention, into their constituent functions and locations. V. Walsh and A. Cowey, "Transcranial Magnetic Stimulation and Cognitive Neuroscience," *Nature Neuroscience Reviews* 1 (2000): 73–79; D. Knoch, "Disruption of Right Prefrontal Cortex by Low-Frequency Repetitive Transcranial Magnetic Stimulation Induces Risk-Taking Behavior," *Journal of Neuroscience* 26, no. 24 (2006): 6469–6472; S. Tassy et al., "Disrupting the Right Prefrontal Cortex Alters Moral Judgment," *Social Cognitive and Affective Neuroscience* 7, no. 3 (2012): 282–288, http://scan.oxfordjournals.org/content/early/2011/04/22/scan.nsr008.full.pdf+html.

26. Pattern analysis is also called multivoxel pattern analysis. See generally Frank Tong and Michael S. Pratte, "Decoding Patterns of Human Brain Activity," *Annual Review of Psychology* 63 (2012): 483–509; and Sebastian Seung, *Connectome: How the Brain's Wiring Makes Us Who We Are* (Boston: Houghton Mifflin Harcourt, 2012), 39–59. The Human Connectome Project, a five-year, $40 million effort funded by the U.S. National Institutes of Health in Bethesda, Maryland, that got under way in 2010, aims to map the human brain's wiring by using a variety of techniques, including fMRI; J. Bardin, "Neuroscience: Making Connections," *Nature* 483 (2012): 394–396. Russell Poldrack describes the Romney example in Miller, "Growing Pains for fMRI," 1414.

27. *Frontline*, "Interview: Deborah Yurgelun Todd," interview on "Inside the Teenage Brain," PBS, January 31, 2002, http://www.pbs.org/wgbh/pages/frontline/shows/teenbrain/interviews/todd.html.

28. David Dobbs, "Fact or Phrenology? Medical Imaging Forces the Debate over Whether the Brain Equals Mind," *Scientific American,* April 2005, http://daviddobbs.net/articles/fact-or-phrenology-medical-imaging-forces-the-debate-over-wh.html; Amanda Schaffer, "Head Case: Roper v. Simmons Asks How Adolescent and Adult Brains Differ," *Slate,* October 15, 2004, http://www.slate.com/articles/health_and_science/medical_examiner/2004/10/head_case.html.

29. To compensate for the relatively slow temporal resolution of fMRI, many researchers combine it with EEG measures and deep electrodes into neurons themselves that capture neuronal activity much closer to real time. The spatial resolution of EEG is inferior to that of fMRI, so they complement each other. A newer technology, magnetoencephalography, surpasses EEG as a measure of neuronal activity; its temporal resolution is comparable with that of electrodes that actually record directly from brain tissue. A relatively new noninvasive method called tensor diffusion imaging can visualize the sweep of large nerve tracts that connect active regions. For an overview, see *Human Functional Brain Imaging, 1990–2009,* 35. Although fMRI has yielded much valuable information, it suffers from two fundamental limitations. First, changes in the blood are slow relative to the speed of neural activity, so it is impossible to measure rapid brain events. Second, the source of the activity can be localized only to the nearest blood vessel; fMRI is not precise enough to provide detailed information about activity in specific neurons and circuits. For a brief overview of attempts to overcome these limitations, see Alan Jasanoff, "Adventures in Neurobioengineering," *ACS Chemical Neuroscience* 3, no. 8 (2012): 575. Among other things, he is devising new

contrast agents whose magnetic properties are altered by events in the neurons themselves, rather than in their surrounding blood vessels. If such an approach is successful, it could reveal an unprecedented level of information about brain activity as it unfolds in real time.

30. In the case of memory encoding, only a modest and sparsely distributed ensemble of neurons may be involved. When researchers destroyed neurons in mice containing high levels of a specific protein within the lateral amygdala, they were able to disrupt auditory fear encoding. Yet ablating adjacent neurons containing lower levels of the protein did not impair encoding. This study suggests that high spatial resolution—perhaps higher than some fMRI machines possess—may be needed to detect small but key subpopulations of neurons within given structures; see Jin-Hee Han et al., "Selective Erasure of a Fear Memory," *Science* 323, no. 5920 (2009): 1492–1496, http://local hopf.cns.nyu.edu/events/spf/SPF_papers/Han%20Josselyn%202009%20Creb%20and%20fear %20memory.pdf. On the practice-suppression effect, see Jason M. Chein and Walter Schneider, "Neuroimaging Studies of Practice-Related Change: fMRI and Meta-analytic Evidence of a Domain-General Control Network for Learning," *Cognitive Brain Research* 25 (2005): 607–623, https://www.ewi-ssl.pitt.edu/psychology/admin/faculty-publications/200702011518450.fMRI.pdf.

31. Pashler quoted in Laura Sanders, "Trawling the Brain: New Findings Raise Questions About Reliability of fMRI as Gauge of Neural Activity," *Science News* 176, no. 13 (2009): 16, http:// laplab.ucsd.edu/news/trawling_the_brain_-_science.pdf. The mischief is not unique to analyzing neuroimaging data. In other disciplines, such as astrophysics and gene mapping, researchers also rely heavily on conceptual and statistical assumptions, but brain-imaging studies happen to be the catnip du jour for the popular press, which too often misrepresents their significance. Craig M. Bennett and Michael B. Miller, "How Reliable Are the Results from Functional Magnetic Resonance Imaging?," *Annals of the New York Academy of Sciences* 1191 (2010): 133–155. Bennett and Miller state that even for the same individual, brain structure and function change over time at the micro and, given enough time, the macro levels. This underlying complexity and plasticity may account for apparent difficulties in reproducing fMRI studies. See also Joshua Carp, "On the Plurality of (Methodological) Worlds: Estimating the Analytic Flexibility of fMRI Experiments," *Frontiers in Neuroscience* 6 (2012): 1–13.

32. Craig M. Bennett et al., "Neural Correlates of Interspecies Perspective Taking in the Post-mortem Atlantic Salmon: An Argument for Multiple Comparisons Correction," *Journal of Serendipitous and Unexpected Results* 1, no. 1 (2010): 1–5, http://prefrontal.org/files/posters/Bennett -Salmon-2009.pdf.

33. Jon Bardin, "The Voodoo That Scientists Do," *Seed*, February 24, 2009, http://seedmagazine .com/content/article/that_voodoo_that_scientists_do/.

34. Edward Vul et al., "Puzzlingly High Correlations in fMRI Studies of Emotion, Personality, and Social Cognition," *Perspectives on Psychological Science* 4, no. 3 (2009): 274–290.

35. When Vul and colleagues' article appeared in hardcopy in May 2009, several lengthy invited commentaries accompanied it.

36. On what brain images are not, see Adina L. Roskies, "Are Neuroimages Like Photographs of the Brain?," *Philosophy of Science* 74 (2007): 860–872; Racine et al., "Brain Imaging"; and A. Bosja and Scott O. Lilienfeld, "College Students' Misconceptions About Abnormal Psychology," poster presented at Undergraduate SIRE Conference, Emory University, April 2010.

37. Eric Racine, Ofek Bar-Ilan, Judy Illes, "fMRI in the Public Eye," *Nature Reviews Neuroscience* 6, no. 2 (2005): 159–164, 160. See also Jean Decety and John Cacioppo, "Frontiers in Human Neuroscience: The Golden Triangle and Beyond," *Perspectives on Psychological Science* 5, no. 6 (2010): 767–771. They write, "The notion seems to be that if you can visualize putative changes in brain activation [in response] to specific tasks, then you have captured something real. . . . There are no more concerns about the validity of self-reports or behavior—if it can be seen in the brain then it must be true. The seductive appeal of neuroimaging is reminiscent of how most people feel about eye-witness evidence, and it is at least as fraught with error" (767). Paul Zak quoted in Eryn

Brown, "The Brain Science Behind Economics," *Los Angeles Times*, March 3, 2012, A13. See Randy Dotinga, "People Love Talking About Themselves, Brain Scans Show," *U.S. News & World Report*, May 7, 2012, http://health.usnews.com/health-news/news/articles/2012/05/07/people-love-talking-about-themselves-brain-scans-show; Mark Thompson, "Study Points at a Clear-Cut Way to Diagnose PTSD," *Time*, January 25, 2010, www.time.com/time/nation/article/0,8599,1956315,00.html; Ian Sample and Polly Curtis, "Hell Hath No Fury Like a Man Scorned, Revenge Tests Reveal," *Guardian,* January 18, 2006, http://www.guardian.co.uk/science/2006/jan/19/research.highereducation. For "neurologism," see Judy Illes, "Neurologisms," *American Journal of Bioethics*, 9, no. 9 (2009): 1.

38. "White House Conference on Early Childhood Development and Learning," April 17, 1997, http://clinton3.nara.gov/WH/New/ECDC/. The Education Commission of the States and the Charles A. Dana Foundation held a similar meeting in 1996 (see Education Commission of the States, "Bridging the Gap Between Neuroscience and Education," September 1996, http://www.ecs.org/clearinghouse/11/98/1198.htm), and the Ethics and Public Policy Center held a more critical conference in 1998 ("Neuroscience and the Human Spirit," Washington, DC, September 24–25, 1998).

39. Nancy C. Andreasen, *The Broken Brain: The Biological Revolution in Psychiatry* (New York: Harper and Row, 1984), 260.

40. Herbert Pardes, "Psychiatric Researchers, Current and Future," *Journal of Clinical Psychopharmacology* 6 (1986): A13–A14, at A13.

41. Thomas R. Insel, "Translating Science into Opportunity," *Archives of General Psychiatry* 66, no. 2 (2009): 128–133.

42. Neely Tucker, "Daniel Amen Is the Most Popular Psychiatrist in America. To Most Researchers and Scientists, That's a Very Bad Thing," *Washington Post*, August 9, 2012, http://articles.washingtonpost.com/2012-08-09/lifestyle/35493561_1_psychiatric-practices-psychiatrist-clinics. Note, however, that SPECT can be helpful in identifying other conditions such as epilepsy, stroke, trauma, and some kinds of dementia. Daniel Carlat, "Brain Scans as Mind Readers: Don't Believe the Hype," *Wired*, May 19, 2008, http://www.wired.com/medtech/health/magazine/16-06/mf_neurohacks?currentPage=all; Martha J. Farah and Seth J. Gillihan, "The Puzzle of Neuroimaging and Psychiatric Diagnosis: Technology and Nosology in an Evolving Discipline," *American Journal of Bioethics—Neuroscience* 3 (2012): 1–11.

43. "People will be very busy easily for the next 20 years," says Peter Bandettini. "I would say that fMRI in many aspects hasn't really even begun." Cited in Kerri Smith, "Brain Imaging: fMRI 2.0," *Nature* 484 (2012): 24–26, at 26.

Chapter 2

1. Martin Lindstrom, *Buyology: Truth and Lies About Why We Buy* (New York: Broadway Books, 2008), 15. Lindstrom followed up with *Brandwashed: Tricks Companies Use to Manipulate Our Minds and Persuade Us to Buy* (New York: Crown Business, 2011). A sample of recent neuromarketing books (some careful and realistic, and others more superficial and slick) includes Erik du Plessis, *The Branded Mind: What Neuroscience Really Tells Us About the Puzzle of the Brain and the Brand* (London: Kogan Page, 2011); Roger Dooley, *Brainfluence: 100 Ways to Persuade and Convince Consumers with Neuromarketing* (Hoboken, NJ: Wiley, 2011); A. K. Pradeep, *The Buying Brain: Secrets for Selling to the Subconscious Mind* (New York: Wiley, 2010); Susan M. Weinschenk, *Neuro Web Design: What Makes Them Click?* (Indianapolis, IN: New Riders Press, 2009); and Patrick Renvoisé and Christophe Morin, *Neuromarketing: Is There a "Buy Button" in the Brain? Selling to the Old Brain for Instant Success* (San Francisco, CA: SalesBrain, 2005). The Lindstrom quotations are from *Buyology*, 11; and Martin Lindstrom, "Our Buyology: The Personal

Coach," http://thepersonalcoach.ca/documents/Buyology_chapter_1(4).pdf. On Lindstrom being named as one of *Time*'s top 100, see "*Time* Top 100, 2009," http://www.martinlindstrom.com/in dex.php/cmsid_buyology_TIME100. Katie Bayne, chief marketing officer of Coca-Cola North America, praised neuromarketing as "provid[ing] you with more natural and unedited responses than you get when you force people through the cognitive loop of having to annunciate how they feel," in Steve McClellan, "Mind over Matter," *Adweek*, February 18, 2008, http://www.adweek .com/news/television/mind-over-matter-94955. See also Rachel Kaufman, "Neuromarketers Get Inside Buyers' Brains," *CNNMoney.com*, March 18, 2010, http://money.cnn.com/2010/03/17 /smallbusiness/neuromarketing/index.htm?section=money_smbusiness; and Joseph Plambeck, "Brain Waves and Newsstands," *New York Times*, September 5, 2010, http://mediadecoder.blogs .nytimes.com/2010/09/05/brain-waves-and-newsstands/.

2. The term "neuromarketing" is widely attributed to Ale Smidts, a marketing professor at Erasmus University in Rotterdam, per Thomas K. Grose, "Marketing: What Makes Us Buy?," *Time*, September 17, 2006. On application of the tools of neuroscience, see Carl Erik Fisher, L. Chin, and Robert Klitzman, "Marketing: Practices and Professional Challenges," *Harvard Review of Psychiatry* 18 (2010): 230–237; Laurie Burkitt, "Neuromarketing: Companies Use Neuroscience for Consumer Insights," *Forbes*, November 16, 2009, http://www.allbusiness.com/marketing-advertis ing/market-research-analysis/13397400-1.html; and Graham Lawton and Clare Wilson, "Mind-Reading Marketers Have Ways of Making You Buy," *New Scientist* 2772 (2010), http://www .newscientist.com/article/mg20727721.300-mindreading-marketers-have-ways-of-making-you -buy.html?page=1. Wanamaker is quoted in Edward L. Lach Jr., "Wanamaker, John," *American National Biography Online*, February 2000, http://www.anb.org/articles/10/10-01706.html. On advertising spending for television, internet, radio, and print, see "Kantar Media Reports U.S. Advertising Expenditures Increased 0.8 Percent in 2011," March 12, 2012, http://www.kantarmedia.com/sites /default/files/press/Kantar_Media_2011_Q4_US_Ad_Spend.pdf. On the failure rate of new products, see Gerald Zaltman, *How Customers Think: Essential Insights into the Mind of the Market* (Boston: Harvard Business Review Press, 2003), 3.

3. Natasha Singer, "Making Ads That Whisper to the Brain," *New York Times*, November 13, 2010, http://www.nytimes.com/2010/11/14/business/14stream.html; Kevin Randall, "Neuromarketing Hope and Hype: 5 Brands Conducting Brain Research," *Fast Company*, September 15, 2009, http://www.fastcompany.com/1357239/neuromarketing-hope-and-hype-5-brands-conduct ing-brain-research. Carl E. Fisher and colleagues reviewed the websites of sixteen neuromarketing firms and reported the following: Thirteen companies "described their methodology, but these descriptions were often insufficient to determine what was being done." The authors concluded that there was "a paucity" of peer-reviewed reports on the company sites. Eleven websites referenced none at all, and only one company "provided citations for its specific claims." Nine of the companies, however, listed staff members with advanced science degrees. See Fisher, Chin, and Klitzman, "Defining Neuromarketing." See also Pradeep, *Buying Brain*. Nielsen acquired Neurofocus in May 2011, prompting insiders to ask, "Is this the start of a stampede for mainstream market research firms & ad agencies to have their own neuromarketing units?" See Roger Dooley, "Nielsen to Acquire Neurofocus," May 20, 2011, http://www.neurosciencemarketing.com/blog/articles/nielsen -to-acquire-neurofocus.htm. FKF presents a vastly oversimplified lesson in the functional anatomy of the brain. "A key part of that data is how the brain reacts in nine well known and well mapped areas, such as the Ventral Striatum (reward), Orbitofrontal Prefrontal Cortex (wanting), Medial Prefrontal Cortex (feeling connected), Anterior Cingulate Cortex (conflict) and the Amygdala (threat/challenge)." See http://www.fkfappliedresearch.com/AboutUs.html. "Neuromarketing is the study of how humans choose, and choice is inescapably a biological process," says David Lewis of Neuroco, a neuromarketing firm in the U.K., in Thomas Mucha, "This Is Your Brain on Advertising," August 1, 2005, http://money.cnn.com/magazines/business2/business2_archive/2005 /08/01/8269671/index.htm.

4. Adam L. Penenberg, "NeuroFocus Uses Neuromarketing to Hack Your Brain," *Fast Company*, August 8, 2011, http://www.fastcompany.com/magazine/158/neuromarketing-intel-paypal. See also Stuart Elliott, "Is the Ad a Success? Brainwaves Tell All," *New York Times*, March 31, 2008; and Nick Carr, "Neuromarketing Could Make Mind Reading the Ad-Man's Ultimate Tool," *Guardian*, April 2, 2008, http://www.guardian.co.uk/technology/2008/apr/03/news.advertising. On the "buy button," see Clint Witchalls, "Pushing the Buy Button," *Newsweek*, March 22, 2004. On Sales-Brain, see "Neuromarketing: Understanding the Buy Buttons in Your Customer's Brain," http://www.salesbrain.com/are-you-delivering-with-impact-on-the-brain/speaking-engagements/. The BrightHouse Institute for Thought Sciences partnered with leading neuroscience professors from Emory University to better understand human thought and apply the knowledge to societal and business concerns. BrightHouse used Emory-owned fMRI scanners to "unlock the consumer mind," as its promotional material put it. BrightHouse Institute for Thought Sciences news release, June 22, 2002, http://www.prweb.com/releases/2002/06/prweb40936.htm. "Imagine being able to observe and quantify a consumer's true response to something without the influence of groupthink and other biases that plague current research approaches," said Brian Hankin, president of Bright-House, quoted in Scott LaFee, "Brain Sales: Through Imaging, Marketers Hope to Peer Inside Consumers' Minds," *San Diego Union Tribune,* July 28, 2004, http://legacy.utsandiego.com/news /.../20040728-9999-lz1c28brain.html.

5. Michael Brammer, "Brain Scam?," *Nature Neuroscience* 7, no. 7 (2004): 683, http://www.nature.com/neuro/journal/v7/n7/pdf/nn0704-683.pdf. Brammer notes that "cognitive scientists, many of whom watched from the sidelines as their molecular colleagues got rich, are now jumping on the commercial bandwagon." On the absence of documentation, see "NeuroStandards Project White Paper," *Advertising Research Foundation NeuroStandards Collaboration Project* 1.0, October 2011, 7, http://neurospire.com/pdfs/arfwhitepaper.pdf. NeuroFocus has Eric Kandel, a 2000 Nobel Prize winner in medicine/physiology, on its advisory board.

6. Lisa Terry, "Learning What Motivates Shoppers (Quarterly Trend Report)," *Advertising Age*, July 25, 2011, 2–19, citing a spring 2011 survey by Greenbook Industry Trends Report. The director general of the European Society for Opinion and Market Research (a global market research association) said in 2011, "As far as [we are] concerned, it is clear that Neuroscience has a growing commercial following and application, but that there also remain a number of key questions, the three main ones [being]: Why is there very little peer-reviewed literature on the topic? Are these methods truly immune from subjectivity and bias? What is the true dollar value of neuroscientific research?" Remarks by ESOMAR director Finn Raben on June 8, 2011, http://rwconnect.esomar.org/2011/06/08/neuroscience-seminar-2011/. The quotation from Roger Dooley is in a personal communication with authors, September 17, 2010, http://www.neurosciencemarketing.com/blog/.

7. Martin Lindstrom, "10 Points Business Leaders Can Learn from Steve Jobs," *Fast Company*, October 15, 2011, http://www.martinlindstrom.com/fast-company-10-points-business-leaders-can-learn-from-steve-jobs/; Martin Lindstrom, "You Love Your iPhone, Literally," *New York Times*, September 30, 2011; Ben R. Newell and David R. Shanks, "Unconscious Influences on Decision Making: A Critical Review," *Behavioral and Brain Sciences* (in press).

8. P. J. Kreshel, "John B. Watson at J. Walter Thompson: The Legitimation of 'Science' in Advertising," *Journal of Advertising* 19, no. 2 (1990): 49–59.

9. Melvin Thomas Copeland, *Principles of Merchandising* (Chicago: A. W. Shaw Company, 1924), 162. On Freudian theory in marketing, see Lawrence R. Samuel, *Freud on Madison Avenue: Motivation Research and Subliminal Advertising in America* (Philadelphia: University of Pennsylvania Press, 2010); and Stephen Fox, *The Mirror Makers* (Urbana: University of Illinois Press, 1997). Fans of *Mad Men* may recall the first episode (July 19, 2007), in which Don Draper, a smoker, is advised by his agency's head of research, a German-accented woman, to deploy the concept of the Freudian death wish to sell Lucky Strike cigarettes. "I find your whole approach perverse," he tells her, repelled by the idea of a death wish, and drops her report in the garbage can. On Dichter, see

"How Ernest Dichter, an Acolyte of Sigmund Freud, Revolutionised Marketing," *Economist*, December 17, 2011, www.economist.com/node/21541706. "His larger philosophy was that a marketer who tries to sell a self-indulgent product must assuage the guilt that goes with it." Morton Hunt, *The History of Psychology* (New York: Doubleday, 1993), 620.

10. Ernest Dichter, *The Strategy of Desire* (Garden City, NY: Doubleday and Company, 1960; repr., New Brunswick, NJ: Transaction Publishers, 2004), 31. Dichter claimed, for example, that smokers liked to use lighters because they fulfilled the human desire to "summon fire . . . the desire for mastery and power. . . . It is also bound up with the idea of sexual potency' " (*Strategy of Desire*, xi). "He sought to scrub the public of a puritanical tradition which, he said, equated consumption, especially of self-indulgent products, with moral transgression"; Daniel Horowitz, *The Anxieties of Affluence* (Amherst: University of Massachusetts Press, 2004), 61. See generally Ernest Dichter, *Handbook of Consumer Motivation: The Psychology of the World of Objects* (New York: McGraw-Hill, 1964); and "How Ernest Dichter, an Acolyte of Sigmund Freud, Revolutionised Marketing." The big emphasis on the "fresh eggs" can be seen in a Betty Crocker commercial from the 1950s at http://www.youtube.com/watch?v=KxdXWw94NgY. Feminist author Betty Friedan would later take Dichter to task for being "paid approximately a million dollars a year for his professional services in manipulating the emotions of American women to serve the needs of business." Betty Friedan, *The Feminine Mystique* (New York: W. W. Norton and Company, 1963; New York: W. W. Norton and Company, 2001), 300. Citations are from the 2001 Tenth Anniversary edition.

11. "Unless all advertising is to become simply a variation on the themes of the Oedipus complex, the death instinct, or toilet training we must recognize that the motives with which we deal should be the manipulable ones," Albert J. Wood, a distinguished Philadelphia businessman and market researcher, told the American Marketing Association in the mid-1950s; cited in Vance Packard, *The Hidden Persuaders* (Philadelphia: D. McKay Company, 1957), 246. See generally Anthony Pratkanis and Elliot Aronson, *Age of Propaganda: The Everyday Use and Abuse of Persuasion* (New York: W. H. Freeman and Co., New York, 1992), 22.

12. In the 1950s, Madison Avenue began to break with its tradition of conformist, follow-the-crowd messaging and targeted particular groups (e.g., young single men, older women, and high-income seniors) for specific products, brands, and services. In differentiating between types of consumer mentality, advertisers gave more sophisticated treatment to the better-educated consumer. "Instead of bludgeoning the customer with razzle-dazzle headlines and ranting copy, admen are buttonholing him with quiet humor, soft talk and attractive art"; "The Sophisticated Sell: Advertisers' Swing to Subtlety," *Time*, September 3, 1956, 68–69, http://www.time.com/time/magazine/article/0,9171,824378,00.html. An influential model for setting advertising objectives and measuring the results was established in 1961. Its acronym was DAGMAR, for Defining Advertising Goals for Measured Advertising Results. The idea was that advertising needed to take a consumer through four levels of progressive understanding, from unawareness to awareness, understanding the product and its benefits, and actually buying the product. See Solomon Dutka and Russell Colley, *DAGMAR: Defining Advertising Goals for Measured Advertising Results* (Lincolnwood, IL: NTC Business Books, 1995). On focus groups, see "Lexicon Valley Takes on Mad Men," in *On the Media,* National Public Radio, June 16, 2012, http://www.onthemedia.org/2012/jun/15/lexicon-valley-takes-mad-men/.

13. Gerald Zaltman, personal communication with authors, October 28, 2010. "A great mismatch exists between the way consumers experience and think about their world and the methods marketers use to collect this information," Zaltman wrote in *How Customers Think*, 37. The Mind of the Market Lab was established in 1997 at Harvard Business School and ended when he retired in 2003. Richard Nisbett and Timothy Wilson, "Telling More Than We Can Know: Verbal Reports on Mental Processes," *Psychological Review 84* (1977): 231–259, is a classic essay on

how people are good at introspecting on the content of their thoughts and desires but are often unable to explain why they think this or want that. As the late advertising magnate David Ogilvy once put it, "People don't think how they feel, they don't say what they think, and they don't do what they say." Sharif Saki, "Market Research and the Primitive Mind of the Consumer," BBC News, March 11, 2006, http://www.bbc.co.uk/news/mobile/business-12581446.

14. Herbert E. Krugman, "Some Applications of Pupil Measurement," *Journal of Marketing Research* 1, no. 4 (1964): 15, 19. The Leo Burnett agency even wired the fingers of a group of housewives to polygraphs to test their reactions to newly filmed TV commercials. Stuart Ewen, "Leo Burnett, Sultan of Sell," *Time*, December 7, 1998. On the use of EEG, see Flemming Hansen, "Hemispheral Lateralization: Implications for Understanding Consumer Behavior," *Journal of Consumer Research* 8, no. 1 (1981): 23–36. Greater electrical asymmetry supposedly reveals affinity or aversion to the product. An increase in left frontal activation suggests a greater affinity for the product and lower scores of right frontal activation indicate not liking a stimulus. Richard J. Davidson, "Affect, Cognition and Hemispheric Specialization," in *Emotions, Cognition and Behavior*, ed. Carroll E. Izard, Jerome Kagan, and Robert B. Zajonc (Cambridge: Cambridge University Press, 1984), 320–365. On steady-state topography, see Max Sutherland, "Neuromarketing: What's It All About?" (originally a talk delivered in February 2007 at Swinburne University in Melbourne), http://www.sutherland.com/Column_pages/Neuromarketing_whats_it_all_about.htm. Anthony Pratkanis, personal communication with authors, May 15, 2012. "Physiological research is not good at predicting success of advertising, and certainly not better than verbal data, though perhaps no worse." Herbert E. Krugman, "A Personal Retrospective on the Use of Physiological Measures of Advertising Response," undated manuscript, ca. 1986, in Edward P. Krugman, *The Selected Works of Herbert E. Krugman: Consumer Behavior and Advertising Involvement* (London: Routledge, 2008), 217.

15. In 1997, Zaltman started the Mind of the Market Laboratory at the Harvard Business School, conducted neuroimaging research with corporate funding, and then shared the results with its sponsors; personal communication to Sally Satel, October 28, 2010. In 2000, Zaltman and colleague, psychologist Stephen Kosslyn, received patent approval to use neuroimaging as a means for validating whether a stimulus such as an advertisement, a communication, or a product evokes a certain mental response, such as emotion, preference, or memory, or to predict the consequences of the stimulus on later behavior, such as consumption or purchasing; http://www.google.com/patents?vid=USPAT6099319. The patent was granted in 2000 and in 2008 was sold to NeuroFocus, when Kosslyn joined the Neurofocus scientific advisory board; see "Neuromarketing Patent Changes Hands," *Neuromarketing*, September 4, 2008, http://www.neurosciencemarketing.com/blog/articles/neuromarketing-patent-changes-hands.htm. On the study, see Gerald Zaltman, *How Customers Think* (Boston: Harvard Business School Press, 2003), 119–121. Zaltman has since shifted his focus to the so-called Zaltman Metaphor Elicitation Technique, an interview protocol that plumbs the unconscious values underlying consumers' reactions to products and marketing campaigns. See the Olson Zaltman Associates website, http://www.olsonzaltman.com/, for more details. "Brain-based research has little deep to say about the meanings and motives that steer individuals' choices," Zaltman told the authors in a personal communication, October 28, 2010.

16. NeuroFocus has created animosity on Madison Avenue, going so far as to ambush the Advertising Research Foundation's release of its findings during its annual conference and releasing its own set of "standards." See http://www.mediapost.com/publications/article/166128/ad-industry-release-final-neuromarketing-report.html#ixzz1jxow4Xap. Ann Parson, "Neuromarketing: Prove Thyself and Protect Consumers," Dana Foundation, December 2011, http://www.dana.org/media/detail.aspx?id=34744. On Kahneman's work, see Daniel Kahneman, *Thinking Fast and Slow* (New York: Farrar, Strauss and Giroux, 2011).

17. Kahneman, *Thinking Fast and Slow*, 278, 367.

18. Kahneman built on the System 1 and 2 distinction originally described by Keith E. Stanovich and Richard F. West; see Stanovich and West, "Individual Differences in Reasoning: Implications for the Rationality Debate," *Behavioral and Brain Sciences* 23, no. 5 (2000): 645–726, http://www.keithstanovich.com/Site/Research_on_Reasoning_files/bbs2000_1.pdf. The blockbuster book by journalist Malcolm Gladwell, *Blink: The Power of Thinking Without Thinking* (New York: Little, Brown and Company, 2005), became the rallying cry to follow one's gut feelings. It eventually inspired a blink backlash; Christopher F. Chabris and Daniel J. Simons, *The Invisible Gorilla—and Other Ways Our Intuition Deceives Us* (New York: Random House, 2010); Wray Herbert, *On Second Thought: Outsmarting Your Mind's Hard-Wired Habits* (New York: Crown, 2010); Daniel Kahneman, "Don't Blink: The Hazards of Confidence," *New York Times Magazine*, October 19, 2011. See the homepage on the Lucid Systems website, http://www.lucidsystems.com/. Says Gemma Calvert, cofounder and managing director of Neurosense, "What you really want to do is look inside the black box and find out what is actually happening in the brain. . . . This technique gets at insights that focus groups can't begin to explain." Calvert is quoted in Eric Pfanner, "On Advertising: Better Ads with MRIs?," *New York Times*, March 26, 2006, http://www.nytimes.com/2006/03/26/business/worldbusiness/26iht-ad27.html?_r=0. "We can use brain imaging to gain insight into the mechanisms behind people's decisions in a way that is often difficult to get at simply by asking a person or watching their behavior," says Dr. Gregory Berns, a psychiatrist at Emory University. Berns is quoted in Alice Park, "The Brain: Marketing to Your Mind," *Time*, January 29, 2007, http://www.time.com/time/magazine/article/0,9171,1580370,00.html#ixzz1h1q7UYIc. "Companies that rely exclusively on traditional measures, focused only at the conscious level, are missing a critical component of what drives purchase behavior," Dr. Carl Marci of Innerscope Research told *Fast Company*. "The vast majority of brain processing (75 to 95%) is done below conscious awareness. Because emotional responses are unconscious, it is virtually impossible for people to fully identify what caused them through conscious measures such as surveys and focus groups." From Jennifer Williams, "Campbell's Soup Neuromarketing Redux: There's Chunks of Real Science in That Recipe," *FastCompany*, February 22, 2010.

19. Stanford's business school has a major in behavioral marketing: http://www.gsb.stanford.edu/phd/fields/marketing/. The MIT Sloan School of Management has a neuroeconomics lab: http://blog.clearadmit.com/2012/04/mit-sloan-researchers-use-neuroscience-to-understand-consumer-spending/. So does the University of California at Berkeley: http://neuroecon.berkeley.edu/. Harvard faculty member Uma Karmarkar has a Ph.D. in neuroscience and marketing: http://drfd.hbs.edu/fit/public/facultyInfo.do?facInfo=bio&facId=588196. Topics in neuroscience include memory; neural correlates of short- and long-term reward; the role of emotion; and the expectation, experience, and recall of value. To a significant degree, all contribute to preference formation, decision making, and the effect of branding. See Hilke Plassmann et al., "What Can Advisers Learn from Neuroscience?," *International Journal of Advertising* 26, no. 2 (2007): 151–175; Antonio Rangel, Colin Camerer, and P. Read Montague, "A Framework for Studying the Neurobiology of Value-Based Decision-Making," *Nature Reviews Neuroscience* 9 (2008): 6; Paul W. Glimcher, Ernst Fehr, Colin Camerer, and Russell A. Poldrack, eds., *Neuroeconomics: Decision-Making and the Brain* (San Diego, CA: Academic Press, 2009); Paul W. Glimcher, *Foundations of Neuroeconomic Analysis* (New York: Oxford University Press, 2011); and Nick Lee, Amanda J. Broderick, and Laura Chamberlain, "What Is 'Neuromarketing'? A Discussion and Agenda for Future Research," *International Journal of Psychophysiology* 63 (2007): 199–204. There is now a neuromarketing textbook: Leon Zurawicki, *Neuromarketing: Exploring the Brain of the Consumer* (Berlin: Springer, 2010). On Plassmann's experiment, see Hilke Plassmann et al., "Marketing Actions Can Modulate Neural Representations of Experienced Utility," *Proceedings of the National Academy of Sciences* 105, no. 3 (2008): 1050–1054.

20. Samuel M. McClure et al., "Neural Correlates of Behavioral Preference for Culturally Familiar Drinks," *Neuron* 44 (2004): 379–387. For what was probably the first taste test, see N. H.

Pronko and J. W. Bowles Jr.,"Identification of Cola Beverages. I. First Study," *Journal of Applied Psychology* 32, no. 3 (1948): 304–312.

21. Notably, in a series of Coke and Pepsi taste tests, they showed that patients with damage specifically involving the ventromedial prefrontal cortex (VMPC), an area important for emotion, did not demonstrate the normal preference bias when they were exposed to brand information. The result that VMPC damage abolishes the "Pepsi paradox" suggests that the VMPC is an important part of the neural substrate for translating commercial images into brand preferences. Michael Koenigs and Daniel Tranel, "Prefrontal Cortex Damage Abolishes Brand-Cued Changes in Cola Preference," *Social Cognitive Affective Neuroscience* 3, no. 1 (2008): 1–6. Montague is quoted in Steve Connor, "Official: Coke Takes Over Parts of the Brain That Pepsi Can't Reach," *Independent*, October 17, 2004, http://labs.vtc.vt.edu/hnl/cache/coke_pepsi_independent_co_uk .htm. Notably, aesthetics can change perception of the product. Coke learned this in the 2011 holiday season. After the company introduced white cans with polar bears, a change from the signature red cans, complaints rolled in about the drink no longer tasting like Coke. Mike Esterl, "A Frosty Reception for Coca-Cola's White Christmas Cans," *Wall Street Journal*, December 1, 2011.

22. Eric Berger, "Coke or Pepsi? It May Not Be up to Taste Buds," *Houston Chronicle*, October 18, 2004; Sandra Blakeslee, "If Your Brain Has a 'Buy Button,' What Pushes It?," *New York Times*, October 19, 2004; Mary Carmichael, "Neuromarketing: Is It Coming to a Lab Near You?," *Frontline PBS*, November 9, 2004, http://www.pbs.org/wgbh/pages/frontline/shows/persuaders/etc /neuro.html; Alok Jha, "Coke or Pepsi? It's All in the Head," *Guardian*, July 29, 2004; Melanie Wells, "In Search of the Buy Button," *Forbes*, September 1, 2003, http://www.forbes.com/forbes/2003 /0901/062.html.

23. Brian Knutson et al., "Neural Predictors of Purchases," *Neuron* 53, no. 1(2007): 147–156; Knutson says, "I believe anticipatory emotions not only bias but drive decision making." Knutson is quoted in Park, "The Brain: Marketing to Your Mind," http://www.time.com/time/magazine /article/0,9171,1580370,00.html#ixzz1h1q7UYIc. The researchers surmised that insula activation reflected the unpleasant prospect of paying too much. In the Knutson experiment, although brain activity elicited during the viewing of a product and the decision to buy were closely matched, a postexperiment assessment that queried subjects on how much they liked the product they bought or whether they thought that it was a good deal turned out to be less closely linked with neural patterns. An important question for the potential of neuromarketing, according to Dan Ariely and Gregory Berns, is "whether the neural signal at the time of, or slightly before, the decision (assumed to be a measure of 'decision utility') can be a good predictor of the pleasure or reward at the time of consumption (the 'experienced utility')." Dan Ariely and Gregory S. Berns, "Neuromarketing: The Hope and the Hype of Neuroimaging in Business," *Nature Reviews Neuroscience* 11 (2010): 284–292, 285. Notably, depending on the statistical analysis used, the self-reported preference was an even better predictor of purchase. Ariely and Berns speculate that someday, neuromarketing approaches might guide the designing of a political candidate's appearance and message content.

24. Gregory S. Berns and Sara E. Moore, "A Neural Predictor of Cultural Popularity," *Journal of Consumer Psychology*, June 8, 2011, http://www.cs.colorado.edu/ . . . /Berns_JCP%20-%20Pop music%20final.pdf. Predicting which songs sell more or less than a threshold are two different numbers, which is why they don't add to 100 percent (similar to false positive and false negative rates in diagnostic tests). G. Berns, personal communication with authors, July 16, 2012.

25. "Neuroco charges an average of $90,000 per study. And its list of services is growing: The firm will evaluate the subliminal power of colors, logos, or product features. It measures the mental might of music or jingles, the heft of celebrity endorsers, and the most brain-wave-soothing designs for store layouts. The company is even testing neurological reactions to smell and touch, and has worked with U.K. auto dealers to gauge responses to the feel of automobile upholstery and the sound of a car door as it slams; see Thomas Mucha, "This Is Your Brain on Advertising," *CNN*

Money, August 1, 2005, http://money.cnn.com/magazines/business2/business2_archive/2005/08/01/8269671/index.htm.

26. McClellan, "Mind over Matter"; Kevin Randall, "The Rise of Neurocinema—How Hollywood Studios Harness Brain Waves to Win Oscars," *Fast Company*, February 25, 2011; Jessica Hamzelou, "Brain Scans Can Predict How You'll React to a Movie Scene," *Gizmodo*, September 9, 2010, http://www.gizmodo.com.au/2010/09/brain-scans-can-predict-how-youll-react-to-a-movie-scene/#more-416708; April Gardner, "Neurocinematics: Your Brain on Film," *NewEnglandFilm.com*, June 30, 2009, http://newenglandfilm.com/magazine/2009/07/neuro.

27. Ellen Byron, "Wash Away Bad Hair Days," *Wall Street Journal*, June 30, 2010.

28. "Product Design and Packaging: Mobile Phone Study," http://www.neurofocus.com/pdfs/Neurofocuscasestudy_ProductDesign.pdf. Left frontal hemisphere activation is correlated with "liking" a stimulus; see R. J. Davidson, "What Does the Prefrontal Cortex 'Do' in Affect? Perspectives on Frontal EEG Asymmetry Research," *Biological Psychology* 67, nos. 1–2 (2004): 219–233. See also G. Vecchiato, "On the Use of EEG or MEG Brain Imaging Tools in Neuromarketing Research," *Computational Intelligence and Neuroscience* 2011, no. 3 (2011), http://www.hindawi.com/journals/cin/2011/643489/. Activation in the posterior frontal cortex may reflect preparation for storage in long-term memory. Rossiter and colleagues used SST to monitor brain waves while people watched TV ads and were able to predict what scenes people would recognize a week later. J. R. Rossiter et al., "Brain-Imaging Detection of Visual Scene in Long-Term Memory for TV Commercials," *Journal of Advertising Research* 41 (2001): 13–21.

29. "NeuroStandards Project White Paper," 7, 34, http://neurospire.com/pdfs/arfwhitepaper.pdf.

30. Burkitt, "Neuromarketing"; D. S. Margulies et al., "Mapping the Functional Connectivity of Anterior Cingulate Cortex," *Neuroimage* 37 (2007): 579–588. For fun, see "The Cingulate Cortex Does Everything," *Annals of Improbable Research* 14, no. 3 (2008): 12–15, http://www-personal.umich.edu/~tmarzull/Cingularity.pdf. (Cingularity is a play on the Singularity, Ray Kurzweil's concept of artificial intelligence; see singularity.com.) In a spoof on the cingulate cortex, neuroscientists mused: "The cingulate cortex is responsible for everything. . . . One implication of this hypothesis is that since more and more researchers will find this brain region attractive, the amount of publications should grow unabated."

31. Marco Iacoboni, "Who Really Won the Super Bowl? The Story of an Instant-Science Experiment," *Edge: The Third Culture*, 2006, http://www.edge.org/3rd_culture/iacoboni06/iacoboni06_index.html. Notably, FKF also rated Super Bowl XLI in 2007, dubbed it, advise, "the year of the amygdala," and predicted that the majority of ads would be "unsuccessful." For "year of the amygdala," see Marcus Yam, "This Is Your Brain on Superbowl Ads," *DailyTech*, February 5, 2007, http://www.dailytech.com/This+is+Your+Brain+on+Super+Bowl+Ads/article5991.htm; for "unsuccessful," see Alice Park, "Brain Scans: How Super Bowl Ads Fumbled," *Time*, February 5, 2007, http://www.fkfappliedresearch.com/media3.html. On the role of the amygdala, see Chiara Cristinzio and Patrik Vuilleumier, "The Role of Amygdala in Emotional and Social Functions: Implications for Temporal Lobe Epilepsy," *Epileptologie* 24 (2007): 78–89, http://labnic.unige.ch/nic/papers/CC_PV_EPI07.pdf On humor in ads, see Madelijn Strick et al., "Humor in Advertisements Enhances Product Liking by Mere Association," *Journal of Experimental Psychology: Applied* 15, no. 1 (2009): 35–45. On ranking Super Bowl ads, see Roger Dooley, "Super Bowl Ads Ranked by Brain Scans," *Neuromarketing*, February 2, 2007, http://www.neurosciencemarketing.com/blog/articles/super-bowl-xli-ads.htm. Comscore reported on actual website traffic during the Super Bowl, comparing the number of website visits for each advertiser in real time. It declared GoDaddy.com, whose ad featured a well-endowed model experiencing yet another wardrobe malfunction, the big winner. GoDaddy's site traffic was up 1,500 percent, and it attracted 439,000 unique visitors. The second-biggest gainer was Budweiser, which saw a 500 percent increase in traffic but ran a much larger number of spots than GoDaddy. See "Super Bowl Ads: GoDaddy Girl 1, Neuroscientists 0," February 17, 2006, http://www.neurosciencemarketing.com/blog/articles

/superbowl-ads-brain-godaddy.htm; and Iacoboni, "Who Really Won the Super Bowl?" Interesting history: GoDaddy ads were censored the previous year for being too raunchy; see http://videos .godaddy.com/superbowl_timeline06.aspx. See also Roxanne Khamsi, "Brain Scans Reveal Power of Super Bowl Adverts," *NewScientist,* February 7, 2006, http://www.newscientist.com/article /dn8691, commenting that the ad for Budweiser beer, which featured a "secret fridge," managed to stir the brain's visual areas only. This might lead one to predict that the ad would be less effective, but, in fact, consumer rankings by the newspaper *USA Today* put this commercial as the "most popular." Finally, not all ads have the same goal. One advertiser might be seeking to establish an aura of prestige around its product, another might want to convince potential buyers that its product's formulation is new and improved over similar products, and still others may be seeking to create name recognition for a new brand or, oppositely, reinforce loyalty for an established brand. Ads for products in a mature market are aimed at inducing established consumers to switch; ads for new products may emphasize information to generate new users. Some ads probably work by directly persuading people to buy something now, while others are geared toward changing customers' brand affinity, which will affect behavior at a later time. There is even a good case to be made that what makes advertising believable are not the ads themselves but the long-haul efforts of sellers to offer reliable quality and thus give people reasons to believe their ads. See "Super Bowl Ads: GoDaddy Girl 1, Neuroscientists 0"; Plassman et al., "What Can Advertisers Learn from Neuroscientists?"; and John E. Calfee, *Fear of Persuasion: A New Perspective on Advertising and Regulation* (Washington, DC: AEI Press, 1997).

32. Calfee, *Fear of Persuasion,* 1.

33. Vance Packard, *Hidden Persuaders.* See also Marshall McLuhan, *The Mechanical Bride: Folklore of Industrial Man* (New York: Vanguard, 1951). From preface: "Ours is the first age in which many thousands of the best-trained individual minds have made it a full-time business to get inside the collective public mind. To get inside in order to manipulate, exploit, control is the object now," at http://home.roadrunner.com/~lifetime/mm-TMB.htm. Packard, *Hidden Persuaders,* 28, 167; "The Hidden Persuaders, by Vance Packard," review of *The Hidden Persuaders,* by Vance Packard, *New Yorker,* May 18, 1957, 167; Nick Johnson, "Review of Vance Packard's *The Hidden Persuaders,*" *Texas Law Review* 36 (1958): 708–715 (molding, 708; Orwell, 713).

34. Randall Rothenberg, "Advertising; Capitalist Eye on the Soviet Consumer," *New York Times,* February 15, 1989. The theory of a robotic Manchurian Candidate was developed by the CIA in 1954. See generally John Marks, *The Search for the Manchurian Candidate: The CIA and Mind Control* (New York: Times Books, 1979). In 1959, Richard Condon published a novel, *The Manchurian Candidate,* about the son of a prominent U.S. political family, a former POW in Korea who is brainwashed to become an unknowing assassin for the Communists. The movie was released in 1962. On the Cincinnati Redlegs, see http://www.sportsecyclopedia.com/nl/cincyreds /reds.html.

35. "Persuaders Get Deeply Hidden Tool: Subliminal Projection," *Advertising Age* 37 (1957): 127. A *New Yorker* correspondent reported: "About fifty members of the press turned up, and we all sat obediently and receptively, if a bit sadly, in our little mortuary chairs, allowing our brains to be softly broken and entered. . . . We have attended many a history-making hoedown in New York, but this jig was the creepiest" ("Talk of the Town," *New Yorker,* September 21, 1957, 33). Herbert Brean, "'Hidden Sell' Technique Is Almost Here: New Subliminal Gimmicks Now Offer Blood, Skulls, and Popcorn to Movie Fans," *Life,* March 31, 1958, 104; Pratkanis and Aronson, *Age of Propaganda,* 199. Gary P. Radford, "Scientific Knowledge and the Twist in the Tail" (paper presented at the forty-second annual conference of the International Communication Association, Miami, Florida, May 21–25, 1992), http://www.theprofessors.net/sublim.html; Kelly B. Crandall, "Invisible Commercials and Hidden Persuaders: James M. Vicary and the Subliminal Advertising Controversy of 1957" (undergraduate honors thesis, University of Florida, 2006), http://plaza.ufl .edu/cyllek/docs/KCrandall_Thesis2006.pdf.

36. Norman Cousins, "Smudging the Subconscious," *Saturday Review*, October 5, 1957, 20; "Ban on Subliminal Ads, Pending FCC Probe, Is Urged," *Advertising Age*, November 11, 1957, 1; Stuart Rogers, "How a Publicity Blitz Created the Myth of Subliminal Advertising," *Public Relations Quarterly* 37, no. 4 (1992): 12–17; "Psychic Hucksterism Stir Calls for Inquiry," *New York Times*, October 6, 1957, 38; Jack Gould, "A State of Mind: Subliminal Advertising, Invisible to Viewer, Stirs Doubt and Debate," *New York Times*, December 8, 1957, D15.

37. "Subliminal Ads Should Cause Little Concern, Psychologists Told," *Washington Post*, September 2, 1958; James B. Twitchell, *Adcult USA: The Triumph of Advertising in American Culture* (New York: Columbia University Press, 1996), 114; Anthony R. Pratkanis, "The Cargo-Cult Science of Subliminal Persuasion," *Skeptical Inquirer* 16, no. 3 (1992), http://www.csicop.org/si/show /cargo-cult_science_of_subliminal_persuasion.

38. F. Danzig, "Subliminal Advertising—Today It's Just Historic Flashback for Researcher Vicary," *Advertising Age*, September 17, 1962, 33, 72, 74; Raymond A. Bauer, "The Limits of Persuasion: The Hidden Persuaders Are Made of Straw," *Harvard Business Review* 36, no. 5 (1958): 105–110, 105. No research has shown an effect of subliminal stimuli on attitudes or purchasing behavior. See reviews by Sheri J. Broyles, "Subliminal Advertising and the Perpetual Popularity of Playing to People's Paranoia," *Journal of Consumer Affairs* 40 (2006): 392–406; Anthony R. Pratkanis and Anthony G. Greenwald, "Recent Perspectives on Unconscious Processing: Still No Marketing Applications," *Psychology and Marketing* 5 (1988): 339–355; and T. E. Moore, "Subliminal Perception: Facts and Fallacies," *Skeptical Inquirer* 16 (1992): 273–281. Nor are subliminal tapes for weight loss, memory improvement, or enhancing self-esteem effective. See generally L. A. Brannon and T. C. Brock, "The Subliminal Persuasion Controversy: Reality, Enduring Fable, and Polonious' Weasel," in *Persuasion: Psychological Insights and Perspectives*, ed. S. Shavitt and T. C. Brock (Needham Heights, MA: Allyn and Bacon, 1994): 279–293; J. Saegert, "Why Marketing Should Quit Giving Subliminal Advertising the Benefit of the Doubt," *Psychology and Marketing* 4 (1987): 107–120; Brandon Randolph-Seng and Robert D. Mather, "Does Subliminal Persuasion Work? It Depends on Your Motivation and Awareness," *Skeptical Inquirer* 33, no. 5 (2009): 49–53; and Joel Cooper and Grant Cooper, "Subliminal Motivation: A Story Revisited," *Journal of Applied Social Psychology* 32, no. 11 (2002): 2213–2227. Also see a report on Daniel Kahneman's plea to replicate findings such as these in Ed Yong, "Nobel Laureate Challenges Psychologists to Clean Up Their Act," *Nature News*, October 3, 2012, http://www.nature.com/news/nobel-laureate -challenges-psychologists-to-clean-up-their-act-1.11535.

39. Scott O. Lilienfeld et al., *A Review of 50 Great Myths of Popular Psychology: Shattering Widespread Misconceptions about Human Behavior* (Hoboken, NJ: Wiley-Blackwell, 2009); Natasha Singer, "Making Ads That Whisper to the Brain," *New York Times*, November 13, 2010; Mark R. Wilson, Jeannie Gaines, and Ronald P. Hill, "Neuromarketing and Consumer Free Will," *Journal of Consumer Affairs* 42, no. 3 (2008): 389–410; "Neuromarketing: Beyond Branding," *Lancet Neurology* 3 (2004): 71; "News Release: Commercial Alert Asks Feds to Investigate Neuromarketing Research at Emory University," December 17, 2003, http://www.commerciala lert.org/issues/culture/neuromarketing/commercial-alert-asks-feds-to-investigate-neuromarket ing-research-at-emory-university. "We Americans may find out sooner than we think. Orwellian is not too strong a term for this prospect"; "Commercial Alert Asks Senate Commerce Committee to Investigate Neuromarketing," July 12, 2004, http://www.commercialalert.org/issues/cul ture/neuromarketing/commercial-alert-asks-senate-commerce-committee-to-investigate -neuromarketing.

40. Complaint and Request for Investigation, submitted by the Center for Digital Democracy, Consumer Action, Consumer Watchdog, and the Praxis Project, October 19, 2011, 2, http://case -studies.digitalads.org/wp-content/uploads/2011/10/complaint.pdf. The FTC is reviewing the complaint according to Jeffrey Chester of the Center for Digital Democracy, personal communication with authors, January 26, 2013. Boire is quoted in Jim Schnabel, "Neuromarketers: The

New Influence Peddlers?," Dana Foundation, March 25, 2008, http://dana.org/news/features/de tail.aspx?id=11686.

41. FCC's *Manual for Broadcasters,* http://www.fcc.gov/guides/public-and-broadcasting-july -2008. Blitz's statement is at "Neuromarketing, Subliminal Messages, and Freedom of Speech," *Neuroethics &Law Blog,* May 14, 2009, http://kolber.typepad.com/ethics_law_blog/2009/05/ neuromarketing-subliminal-messages-and-freedom-of-speech-blitz.html#comments. As political philosopher Thomas M. Scanlon recognized decades ago, subliminal speech is by no means unique in operating below the radar of our processes of rational reflection. It "happens all the time" that we are influenced by stimuli "without being aware of that influence," and such unconscious changes result not only from hidden stimuli but also from "what is clearly seen and heard." Thomas M. Scanlon, "Freedom of Expression and Categories of Expression," *University of Pittsburgh Law Review* 40 (1979): 519, 525. In light of the uncertainty about how effective neuromarketing might become in the future, some neuroethicists have proposed that researchers and companies using neuromarketing techniques adopt a code of ethics "to ensure beneficent and non-harmful use of the technology." Emily R. Murphy, Judy Illes, and Peter B. Reiner, "Neuroethics of Neuromarketing," *Journal of Consumer Behavior* 7 (2008): 292–302, 292.

42. On the influence of shoppers' moods, see John A. Bargh, "Losing Consciousness: Automatic Influences on Consumer Judgment, Behavior and Motivation," *Journal of Consumer Research* 29 (2002): 280–285; and Mirja Hubert and Peter Kenning, "A Current Overview of Consumer Neuroscience," *Journal of Consumer Behavior* 7 (2008): 272–292. On the influence of background music, see R. E. Milliman, "Using Background Music to Affect the Behavior of Supermarket Shoppers," *Journal of Marketing* 46, no. 3 (1982): 86–91. More generally, see Aradhna Krishna, "An Integrative Review of Sensory Marketing: Engaging the Senses to Affect Perception, Judgment and Behavior," *Journal of Consumer Psychology* 22, no. 3 (2011): 332–351, http://www.sciencedirect.com /science/article/pii/S1057740811000830. On levels of arousal, see D. M. Sanbonmatsu and F. R. Kardes, "The Effects of Physiological Arousal on Information Processing and Persuasion," *Journal of Consumer Research* 15 (1988): 379–385; Michel Tuan Pham, "Cue Representation and Selection Effects of Arousal in Persuasion," *Journal of Consumer Research* 22 (1996): 373–387; R. E. Petty and D. T. Wegener, "Attitude Change: Multiple Roles for Persuasion Variables," in *The Handbook of Social Psychology,* ed. D. Gilbert, S. Fiske, and G. Lindzey, 4th ed. (New York: McGraw-Hill, 1998), 323–390. When customers feel tense, one study found, they are more susceptible to a proven persuasive tactic called "social proof," which consists of an appeal to a product's popularity or best-selling status. R. B. Cialdini and N. J. Goldstein, "Social Influence: Compliance and Conformity," *Annual Review of Psychology* 55 (2004): 591–621. Another study found that when subjects are in a romantic frame of mind, they are more attracted to products that are advertised as hard to get, rare, or "limited edition." Vladas Griskevicius et al., "Fear and Loving in Las Vegas: Evolution, Emotion, and Persuasion," *Journal of Marketing Research* 46 (June 2009): 384–395. See also Sabrina Bruyneel et al., "Repeated Choosing Increases Susceptibility to Affective Product Features," *International Journal of Research in Marketing* 23 (2006): 215–225; and Jing Wang et al., "Trade-offs and Depletion in Choice," *Journal of Marketing Research* 47 (2010): 910–919. These findings are consistent with a persuasion theory called ELM (elaboration likelihood model). This model posits that people find superficial aspects of a message and heuristics persuasive when their motivation or ability for argument processing is low (the "peripheral" route to persuasion). Conversely, when their engagement is high, superficial features of a message and heuristics have less effect on their attitudes. In this instance, the so-called central route to persuasion is activated, and people use careful scrutiny to determine the merits of a message or argument. See generally Richard E. Petty and John T. Cacioppo, *Communication and Persuasion: Central and Peripheral Routes to Attitude Change* (Berlin: Springer-Verlag, 1986).

43. Mya Frazier, "Hidden Persuasion or Junk Science," *Advertising Age,* September 10, 2007, http:// adage.com/article/news/hidden-persuasion-junk-science/120335/. A. S. C. Ehrenberg, "Repetitive

Advertising and the Consumer," *Journal of Advertising Research* 1 (1982): 70–79, 70. "It is not yet whether neuroimaging provides better data than other marketing methods, but through the use of MVPA methods it might be possible to reveal the 'holy grail' of hidden information"; Ariely and Berns, "Neuromarketing," 287.

44. Craig Bennett, "The Seven Sins of Neuromarketing," April 22, 2011, http://prefrontal.org /blog/2011/04/the-seven-sins-of-neuromarketing/. The typical cost of neuromarketing research ranges from about $30 million to $100 million (with Super Bowl spots being sold for an average $2.6 to $2.7 million), as cited in Rachel Kauffman, "Neuromarketers Get Inside Buyers' Brains," *CNNMoney.com*, March 18, 2010, http://money.cnn.com/2010/03/17/smallbusiness/neuromar keting/index.htm?section=money_smbusiness. According to Burkitt, "A marketer can hook 30 consumers up to an EEG device for $50,000. An MRI trial with 20 people would cost more like $40,000," in Burkitt, "Neuromarketing."

45. "NeuroStandards Project White Paper," 7, 30.

46. On Starch, see Sean Brierley, *The Advertising Handbook* (London: Routledge, 1995), 182. Indeed, most marketers still recommend focus groups as useful for obtaining feedback on how consumers view a product. "Far too often a single cutting-edge technology is deployed into projects for technology's sake, without recognizing the importance of cross-method support/validation." Matt Tullman, CEO of Merchant Mechanics, Inc., comment on Roger Dooley, "Your Brain on Soup," *Neuromarketing: Where Brain Science and Marketing Meet* (blog), February 20, 2010, http://www.neurosciencemarketing.com/blog/articles/your-brain-on-soup.htm.

Chapter 3

1. Investigators found that 45 percent of soldiers tried opium, heroin, or both in Vietnam, mostly by smoking. Twenty percent of all enlisted men claimed that they became addicted in Vietnam, and 10.5 percent of 13,760 army enlisted returnees had positive urine tests for barbiturates, opiates, or amphetamines in September 1971, the month of their DEROS (date eligible for return from overseas). See Lee N. Robins, "Vietnam Veterans' Rapid Recovery from Heroin: A Fluke or Normal Expectation?" *Addiction* 88 (1993): 1041–1054, 1046; Rumi Kato Price, Nathan K. Risk, and Edward L. Spitznagel, "Remission from Drug Abuse over a 25-Year Period: Patterns of Remission and Treatment Use," *American Journal of Public Health* 91, no. 7 (2001): 1107–1113. The *New York Times* article is Alvin M. Schuster, "G.I. Heroin Addiction Epidemic in Vietnam," *New York Times*, May 16, 1971, A1. The drug-testing policy was announced in June 1971. In June 1971, under the direction of Jerome Jaffe of the Special Action Office on Drug Abuse Prevention, Lee Robins of Washington University in St. Louis undertook an analysis of the program. For a detailed account of Jaffe's work with the soldiers, see "Oral History Interviews with Substance Abuse Researchers: Jerry Jaffe," Record 16, University of Michigan Substance Abuse Research Center, January 2007, http://sitemaker.umich.edu/substance.abuse.history/oral_history_interviews.

2. Operation Golden Flow is a military term for a routine urinalysis test. Throughout the fall of 1971, the number of positive tests decreased. By February 1972, the positive test rate fell to under 2 percent, at which point the administration declared the "epidemic" under control. Michael Massing, *The Fix: Solving the Nation's Drug Problem* (New York: Simon and Schuster, 1998), 86–131. For Robins's evaluation, see Lee N. Robins, John E. Helzer, and Darlene H. Davis, "Narcotic Use in Southeast Asia and Afterward," *Archives of General Psychiatry* 32 (1975): 955–961.

3. Within the 12 percent of veterans who relapsed was a subgroup of addicts who resumed use for several months but then stopped—that is, not all those who resumed use did so continuously over the entire three-year period. Robins, "Vietnam Veterans' Rapid Recovery from Heroin," 1041–1054, 1046. In a twenty-five-year longitudinal study conducted in 1996–1997, between 5.1 percent and 9.1 percent had used opiates five or more times since military discharge (it is not clear

whether these individuals were part of the 12 percent reported in Robins's initial 1974 article or how many of them became readdicted as opposed to nonaddicted users); Price, Risk, and Spitznagel, "Remission from Drug Abuse." "Revolutionary": Dr. Robert DuPont, the first head of NIDA, personal communication with authors, 1972. "Path-breaking" used by authors Robert Granfield and William Cloud, *Coming Clean: Overcoming Addiction Without Treatment* (New York: New York University Press, 1999), 215. Just over 16 percent of veterans (a subject pool that included those who were positive for opiates when leaving Vietnam and those who were negative plus a subsample of age-matched nonveterans) fulfilled the diagnostic criteria, according to the *Diagnostic and Statistical Manual for Mental Disorders IV*, for drug abuse or dependence at any point between 1972 and 1996. See Price, Risk, and Spitznagel, "Remission from Drug Abuse." Dependence on all substances was at its highest in 1971, 45.1 percent, when opiates were abundant in Vietnam. The rate decreased from 16.4 percent in 1972 to 5.9 percent by 1996. Rumi Kato Price et al., "Posttraumatic Stress Disorder, Drug Dependence, and Suicidality Among Male Vietnam Veterans with a History of Heavy Drug Use," *Drug and Alcohol Dependence* 76 (2004): S31–S43.

4. Alan I. Leshner, "Addiction Is a Brain Disease, and It Matters," *Science* 278, no. 5335 (1997): 45–47. Leshner later stated, "The majority of the biomedical community now considers addiction, in its essence, to be a brain disease." Alan I. Leshner, "Addiction Is a Brain Disease," *Issues in Science and Technology* (2001), http://www.issues.org/17.3/leshner.htm. "The fact that addictions are complex, chronic, and often relapsing brain diseases that need to be managed as intensively as diabetes is taken as a given by the experts," writes the editor in chief of *Health Affairs*; Susan Dentzer, "Substance Abuse and Other Substantive Matters," *Health Affairs* 30, no. 8 (2011): 1398. C. Everett Koop, the former surgeon general, asserts that the brain-disease model of addiction "is recognized by all leading medical and health authorities." C. Everett Koop, "Drug Addiction in America: Challenges and Opportunities," in *Addiction: Science and Treatment for the Twenty-First Century*, ed. Jack E. Henningfield, Patricia B. Santora, and Warren Bickel (Baltimore: Johns Hopkins University Press, 2007), 13. In 2007, then senator Joseph Biden introduced the Recognizing Addiction as a Disease Act of 2007, which stipulated that "addiction is a chronic, relapsing brain disease." The bill, which aimed to merge NIDA with the National Institute on Alcohol Abuse and Alcoholism, had a House companion but died in committee. See U.S. Congress, Senate, Recognizing Addiction as a Disease Act of 2007, S 1101, 110th Cong., 1st Sess. (2007–2008), http://thomas.loc.gov. Reacting to the bill, Daniel Guarnera, government relations liaison for the NAADAC, the Association for Addiction Professionals, said, "NIDA and its scientists have demonstrated overwhelmingly that addiction is not a behavioral trait, but rather is caused by physiological changes to the body that make people want to use addictive substances. This bill allows the terminology to catch up with the science." Quoted in Philip Smith, "Is Addiction a Brain Disease? Biden Bill to Define It as Such Is Moving on Capitol Hill," *Drug War Chronicle*, August 9, 2007, http://stopthedrugwar.org/chronicle/2007/aug/09/feature_addiction_brain_disease. See also Sally Satel and Scott O. Lilienfeld, "Medical Misnomer: Addiction Isn't a Brain Disease, Congress," *Slate*, July 25, 2007, http://www.slate.com/articles/health_and_science/medical_examiner/2007/07/medical_misnomer.html. On high-school antidrug lectures, see Lori Whitten, "NIH Develops High School Curriculum Supplement on Addiction," *NIDA Notes* 16, no. 1 (2001), http://archives.drugabuse.gov/NIDA_Notes/NNVol16N1/NIH.html. Dr. James W. West, the physician director of the Betty Ford Center, says there has been no change in how the center views addiction: "It is now clearly recognized as a brain disease, a bio-psycho-spiritual-socio disease. Now these facts are more widely known and accepted by the public." "BFC Pioneer Dr. James West, 93, Stays the Course," Betty Ford Center, March 1, 2007, http://www.bettyfordcenter.org/news/innews/narticle.php?id=19. At the Washington, D.C., clinic where coauthor Sally Satel works, Partners in Drug Abuse and Rehabilitation, the medical director tells patients that they have a brain disease in his orientation lecture. For the American Society of Addiction Medicine's definition, see American Society of Addiction Medicine, "Public Policy Statement: Definition of Addiction," adopted April 12, 2011,

http://www.asam.org/advocacy/find-a-policy-statement/view-policy-statement/public-policy
-statements/2011/12/15/the-definition-of-addiction. Drug czars: "Today the brain-disease model is
widely accepted in the addiction field, and Barry R. McCaffrey, the White House drug adviser,
routinely invokes it." Michael Massing, "Seeing Drugs as a Choice or as a Brain Anomaly," *New
York Times*, June 24, 2000, http://www.nytimes.com/2000/06/24/arts/seeing-drugs-as-a-choice-or
-as-a-brain-anomaly.html?pagewanted=all&src=pm. John Walters has stated that "addiction is a
disease of the brain." Quynh-Giang Tran, "Drug Policy Chief Looks to the Root of Addiction: U.S.
Eyes 10% Reduction in Abuse in Two Years," *Boston Globe*, July 10, 2002, A3. In remarks on the
nomination of Gil Kerlikowske as director of the Office of National Drug Control Policy, Vice
President Joseph Biden notes, "An addiction is a disease—as Pat Moynihan used to say, disease of
the brain." Remarks on the Nomination of Gil Kerlikowske, Office of the Vice President, March
11, 2009, the White House, http://www.whitehouse.gov/the_press_office/Remarks-of-the-Vice
-President-and-Chief-Kerlikowske-on-his-Nomination-as-the-new-Director-of-the-Office-of-Na
tional-Drug-Control-Policy/. See also National Council on Alcoholism and Drug Dependence, "Addiction Is a Disease, Not a Moral Failure: Kerlikowske," June 12, 2012, http://www.ncadd.org/index
.php/in-the-news/365-addiction-is-a-disease-not-a-moral-failure-kerlikowske. For media attention
to the brain-disease model, see *Addiction*, DVD, produced by John Hoffman and Susan Froemke
(HBO, 2007); Tim Russert and Bill Moyers, "Bill Moyers, Journalist, Discusses His Upcoming PBS
Special on Drug Abuse and Addiction," *Meet the Press*, NBC News, aired March 29, 1998, transcript, LexisNexis; Charlie Rose and Nora Volkow, "The Charlie Rose Brain Series, Year 2," *The
Charlie Rose Show*, PBS, aired August 13, 2012, transcript, LexisNexis; Dick Wolf and Dawn
DeNoon, "Hammered," *Law and Order: Special Victims Unit*, NBC (New York: NBC, October
14, 2009); Drew Pinsky, "Addiction: Do You Need Help?," WebMD Live Events Transcript, Medi-
cineNet.com, November 6, 2003, http://www.medicinenet.com/script/main/art.asp?articlekey=
54633; *Strictly Dr. Drew: Addictions A–Z*, DVD, directed by Christopher Bavelles and José Colomer
(Silver Spring, MD: Discovery Health, 2006); Michael D. Lemonick, "How We Get Addicted,"
Time, July 5, 2007, http://www.time.com/time/magazine/article/0,9171,1640436,00.html; and Je-
neen Interlandi, "What Addicts Need," *Newsweek*, February 23, 2008, http://www.newsweek.com
/2008/02/23/what-addicts-need.html.

   5. "There is an age-old debate over alcoholism: Is the problem in the sufferer's head—something
that can be overcome through willpower, spirituality or talk therapy, perhaps—or is it a physical
disease, one that needs continuing medical treatment in much the same way as, say, diabetes or epilepsy?" D. Quenqua, "Rethinking Addiction's Roots, and Its Treatment," *New York Times*, July 10,
2011, A1.

   6. There is no formal definition of the term "brain disease." Typically, medical professionals use
the term to denote a disruption in functioning of neural processes that (1) is primary, such as those
found in Parkinson's disease, a brain tumor, schizophrenia, autism, or multiple sclerosis (that is,
conditions that are not the result of deliberate behavior, such as drug use; for addiction to be primary, the neural changes need to be the cause of repeated drug use rather than the consequence of
it) or (2) cannot be reversed through behavioral means. Under this framework, smoking is not a
brain disease, but lung cancer brought on by smoking is a lung disease, not because the lung cells
began to divide primarily ("on their own")—indeed, they were prompted to do so under the influence of years of exposure to cigarette smoke—but because, once the malignancy has developed,
the course of the disease cannot be arrested by patient-initiated behavior changes. Remission of
cancer requires surgery, radiation, chemotherapy, and other interventions. The same applies to cir-
rhosis of the liver brought on by alcoholism—that is, once the liver disease has developed, it is
autonomous. Addiction per se meets neither criterion 1 nor 2. Notably, not all primary brain diseases, such as schizophrenia, require only medication. Once the most extreme symptoms are controlled with medication, changes in lifestyle (socialization, psychotherapy) are important. In fact,
once symptoms have been stable for a while, medications can be reduced and, in some patients,

discontinued if the patient becomes adept at limiting stress. For an overview of addiction neurobiology, see Alfred J. Robison and Eric J. Nestler, "Transcriptional and Epigenetic Mechanisms of Addiction," *Nature Reviews Neuroscience* 12 (2011): 623–637. See also Steven E. Hyman, "The Neurobiology of Addiction: Implications for Voluntary Control of Behaviour," in *The Oxford Handbook of Neuroethics*, ed. Judy Illes and B. J. Sahakian (Oxford: Oxford University Press, 2011), 203–218.

7. Kristina Fiore, "Doctor's Orders: Brain's Wiring Makes Change Hard," MedPage Today, January 30, 2010, http://www.medpagetoday.com/Psychiatry/Addictions/18207; Leshner, "Addiction Is a Brain Disease, and It Matters," 46. "Flipping a neurochemical 'switch' " is another common metaphor. See, e.g., Jim Schnabel, "Flipping the Addiction Switch," Dana Foundation, August 26, 2008, http://www.dana.org/news/features/detail.aspx?id=13120. Senator Mike Enzi (R-WY), the ranking minority member on the Senate Committee on Health, Education, Labor and Pensions, has also used the "switch" metaphor: "Science shows us the addiction to alcohol or any other drug is a disease. . . . While the initial decision to use drugs is a choice, there comes a time when continued use turns on the addiction switch in the brain." "Enzi Says HELP Committee Approves Bill to Recognize Addiction as a Disease," press release, Office of U.S. Senator Mike Enzi, June 27, 2007, http://help.senate.gov/old_site/Min_press/2007_06_27_d.pdf. Leshner is quoted at "Fighting Addiction," February 4, 2001, http://www.prnewswire.com/news-releases/cover-fighting-addiction-71196322.html. See also "It may be no more a matter of personal choice to abstain from tobacco than to reverse metastasizing lung cells," in Jack E. Henningfield, Leslie M. Schuh, and Murray E. Jarvik, "Pathophysiology of Tobacco Dependence," in *Neuropsychopharmacology—The Fourth Generation of Progress*, ed. Floyd E. Bloom and David J. Kupfer et al. (New York: Raven Press, 1995), 1715–1729, at 1715.

8. Anna Rose Childress, "Prelude to Passion: Limbic Activation by 'Unseen' Drug and Sexual Cues," *PLoS One* 3 (2008): e1506. Even barely detectable sensory cues, like the faint smell of drugs in the air, can provoke craving.

9. Rita Z. Goldstein and Nora D. Volkow, "Drug Addiction and Its Underlying Neurobiological Basis: Neuroimaging Evidence for the Involvement of the Frontal Cortex," *American Journal of Psychiatry* 159, no. 10 (2002): 1642–1652.

10. Edward Preble and John J. Casey, "Taking Care of Business: The Heroin User's Life on the Street," *International Journal of the Addictions* 4, no. 1 (1969): 1–24. See also Bill Hanson, *Life with Heroin: Voices from the Inner City* (Lexington, MA: Lexington Books, 1985); Charles E. Faupel and Carl B. Klockars, "Drugs-Crime Connections: Elaborations from the Life Histories of Hard-Core Heroin Addicts," *Social Problems* 34, no. 1 (1987): 54–68; and Michael Agar, *Ripping and Running: Formal Ethnography of Urban Heroin Addicts* (Napier, NZ: Seminar Press, 1973). On cocaine addicts, see Philippe Bourgois, *In Search of Respect: Selling Crack in El Barrio* (Cambridge: Cambridge University Press, 2002).

11. Guy Gugliotta, "Revolutionary Thinker: Trotsky's Great-Granddaughter Is Following Her Own Path to Greatness," *Washington Post*, August 21, 2003, C1; Gene M. Heyman, personal communication with authors, September 20, 2012.

12. "Decision-making, ambivalence and conflict [are] central features of the addict's behaviour and experience," writes Nick Heather. Heather, "A Conceptual Framework for Explaining Addiction," *Journal of Psychopharmacology* 12 (1998): 3–7, at 3. See also Jon Elster, "Rational Choice History: A Case of Excessive Ambition," *American Political Science Review* 94, no. 3 (2000): 685–695.

13. William S. Burroughs, *Naked Lunch: The Restored Text*, ed. James Grauerholz and Barry Miles (New York: Grove/Atlantic, 2001): 199. British psychologist Robert West called these "conversion-type experiences." Although they are often triggered by "seemingly trivial events," he writes, "they occur within a system where hidden tensions have been developing all along." Personal communication with authors, July 27, 2005. Christopher K. Lawford, *Moments of Clarity:*

Voices from the Front Lines of Addiction and Recovery (New York: HarperCollins, 2009). For Rolling Stones drummer Keith Richards (who used heroin from his mid-twenties to his mid-thirties), the thought of kicking dope in jail was intolerable: "There was already a permanent black cloud of expecting the shit to hit the fan. I'm facing three charges: trafficking, possession and importing. I'm going to be doing some hard fucking time. I'd better get ready." Keith Richards and James Fox, *Life* (New York: Little, Brown and Company, 2010), 408. The quotation from the recovered alcoholic is in Jim Atkinson, "Act of Faith," *New York Times*, January 26, 2009, http://proof.blogs.nytimes.com/2009/01/26/act-of-faith/. "The rejection of moral considerations in addiction deprives us of our most powerful weapons against addiction," writes Stanton Peele in "A Moral Vision of Addiction: How People's Values Determine Whether They Become and Remain Addicts," *Journal of Drug Issues* 17, no. 2 (1987): 187–215, at 215. He identifies those weapons as "namely the values of both addicted individuals and of the larger society." Personal communication with authors, May 24, 2012.

14. Gene M. Heyman, *Addiction: A Disorder of Choice* (Cambridge, MA: Harvard University Press, 2009), 67–83. According to Heyman, between 77 and 86 percent of individuals who were addicted to drugs or alcohol (substance dependent) at some point in their lives were still without substance problems during the year before the survey according to surveys that distinguish abuse and dependence (e.g., the National Comorbidity Survey, sponsored by the National Institute of Mental Health, and the National Epidemiologic Survey on Alcohol and Related Conditions, sponsored by the National Institute on Alcohol Abuse and Alcoholism). By comparing the number of survey respondents who satisfied criteria for having been addicted at least once in their lifetime (but not within the last year) with the number addicted at the time of the survey, Heyman concluded that between 60 and 80 percent of people who were addicted in their teens and twenties were no longer heavy, problem users by their thirties. Further, they largely maintained their substance-free status over subsequent decades. Note also that the high percentage of remission was not artificially increased by people using marijuana (see table 4.6 on page 81 of Heyman's book). Another study, focusing solely on the National Epidemiologic Survey on Alcohol and Related Conditions (using data from first wave, 2001–2002, as did Heyman), estimated that the lifetime probability of remission was 83.7 percent for nicotine, 90.6 percent for alcohol, 97.2 percent for cannabis, and 99.2 percent for cocaine. These findings demonstrate that in a large nationally representative sample of U.S. adults, the vast majority of individuals experiencing dependence on one or more of these four substances will remit at some time in their lives. Catalina Lopez-Quintero et al., "Probability and Predictors of Remission from Life-time Nicotine, Alcohol, Cannabis or Cocaine Dependence: Results from the National Epidemiologic Survey on Alcohol and Related Conditions," *Addiction* 106 (2011): 657–669.

15. See Wilson M. Compton et al., "Prevalence, Correlates, Disability, and Comorbidity of DSM-IV Drug Abuse and Dependence in the United States: Results from the National Epidemiologic Survey on Alcohol and Related Conditions," *Archives of General Psychiatry* 64 (2007): 566–576 (twelve-month drug dependence remained positively and significantly related to substance-use disorders and each specific mood disorder [except bipolar II disorder], generalized anxiety, and antisocial personality disorder). For the NIDA estimates, see National Institute on Drug Abuse, National Institutes of Health, U.S. Department of Health and Human Services, *Principles of Drug Addiction Treatment: A Research-Based Guide,* 2nd ed. (Rockville, MD: National Institute on Drug Abuse, National Institutes of Health, 1999, rev. 2009), 11.

16. Patricia Cohen and Jacob Cohen, "The Clinician's Illusion," *Archives of General Psychiatry* 14, no. 12 (1984): 1178–1182. It's no coincidence that Lee Robins's results regarding the Vietnam veterans were rejected by many in the addiction research field who dealt mainly with institutionalized addicts. As she wrote in 1993, the "research community . . . resisted giving up the beliefs that heroin was a uniquely dangerous drug, to which a user became addicted very quickly, and [sic] addiction to which was virtually incurable." Robins, "Vietnam Veterans' Rapid Recovery from Heroin Addiction," 1047.

17. Speaking more generally about addiction before the model of the chronic and relapsing brain disease became prevalent, psychologist Thomas Babor wrote, "The primary issue with regard to definitions of [addiction] thus becomes who or which group controls the defining process, and how they use definitions to promote their own ends, be they medical, legal, scientific, or moral"; Thomas F. Babor, "Social, Scientific, and Medical Issues in the Definition of Alcohol and Drug Dependence," in *The Nature of Drug Dependence*, ed. Griffith Edwards and Malcolm Lader (Oxford: Oxford University Press, 1990): 19–36, 33. Alan I. Leshner, personal communication with author, December 6, 2009, "Congressional and NIMH staffers agree that the images, beyond the words, carried great weight in the political discourse that ultimately led to the first mental health parity act in the United States, in 2008," in Bruce R. Rosen and Robert L. Savoy, "fMRI at 20: Has It Changed the World?," *NeuroImage* 62, no. 5 (2012): 1316–1324.

18. "Oral History Interviews with Substance Abuse Researchers: C. Robert 'Bob' Schuster," Record 36, University of Michigan Substance Abuse Research Center, June 14, 2007, http://sitemaker .umich.edu/substance.abuse.history/oral_history_interviews. Schuster further explained in the interview, "Ultimately, however, I believe that not just in this science [but] in all sciences, that a reductionistic approach can only be so successful. As one moves up levels of integration, at these higher levels of integration, new phenomena emerge that are not reducible to those that are below because of the interaction of so many variables. Ultimately, no matter how well we understand the enzymatic and protein pathways of this that or the other thing in the cell, we've got to explain the behavior of the intact, integrated organism. The behavior is always right. It is ultimately the job of the biologist to be able to predict that behavior because the behavior is the reality. The behavioral effects of drugs are the reality. Understanding them at different levels is fine, and I'm all in favor of it. I'm excited about it, but we cannot forget that the end product is that we're trying to change people's behavior." He went on to say that a more technical vocabulary has helped to raise the status of addiction researchers. "Everybody thinks they're an expert in behavior," but only the research elite can speak in biological terms. The low status of addiction research is also represented in a report from the Institute of Medicine of the National Academy of Sciences: "The stigma associated with drug addiction has directly deterred young investigators who might otherwise be interested in pursuing careers in addiction research and treatment. . . . As a result of stigma, the realities of studying often difficult and sometimes frightening patients, and the lack of public funding and support, addiction research is often an undervalued area of inquiry with low visibility, and many scientists and clinicians choose other disciplines in which to develop their careers." Committee to Identify Strategies to Raise the Profile of Substance Abuse and Alcoholism Research, Institute of Medicine, *Dispelling the Myths About Addiction: Strategies to Increase Understanding and Strengthen Research* (Washington, DC: National Academies Press, 1997), 140. "Oral History Interviews with Substance Abuse Researchers: Robert Balster," Record 3, University of Michigan Substance Abuse Research Center, June 2004, http://sitemaker.umich.edu/substance.abuse.history /oral_history_interviews.

19. About 66 percent of the NIDA research budget is directed at basic clinical neuroscience and behavioral research and medication development, while the remainder supports clinical trials of medication and behavioral therapies and most epidemiology and prevention; http://www.drugabuse .gov/about-nida/legislative-activities/budget-information/fiscal-year-2013-budget-information /budget-authority-by-activity-table. For a narrative of established priorities, see "NIDA's Priorities in Tough Fiscal Times," Messages from the Director, NIDA online newsletter, February 2012, http://www.drugabuse.gov/about-nida/directors-page/messages-director/2012/02/nida<#213>s -funding-priorities-in-tough-fiscal-times-flavor of priorities. For the percentage of research funded by NIDA, see "NIDA Funds More Than 85 Percent of the World's Research on Drug Abuse," at "Policy and Research," Office of National Drug Control Policy, the White House, http://www .whitehouse.gov/ondcp/policy-and-research. Jaffe's comments are in "Oral History Interviews with Substance Abuse Researchers: Jerry Jaffe."

20. On relief, see Benjamin Goldstein and Francine Rosselli, "Etiological Paradigms of Depression: The Relationship Between Perceived Causes, Empowerment, Treatment Preferences, and Stigma," *Journal of Mental Health* 12 (2003): 551–563; and Jo C. Phelan, R. Cruz-Rojas, and M. Reiff, "Genes and Stigma: The Connection Between Perceived Genetic Etiology and Attitudes and Beliefs About Mental Illness," *Psychiatric Rehabilitation Skills* 6 (2002): 159–185. On guilt reduction, see Judy Illes et al., "In the Mind's Eye: Provider and Patient Attitudes on Functional Brain Imaging," *Journal of Psychiatric Research* 43 (2008): 107–114; and Emily Borgelt, Daniel Z. Buchman, and Judy Illes, "*This* is Why You've Been Suffering: Reflections of Providers on Neuroimaging in Mental Health Care," *Journal of Bioethical Inquiry* 8, no. 1 (2011): 15–25. On the value of images, see Daniel Z. Buchman et al., "Neurobiological Narratives: Experiences of Mood Disorder Through the Lens of Neuroimaging," *Sociology of Health and Illness* 35, no. 1 (2013): 66–81.

21. See the statement of the American Society for Addiction Medicine, December 15, 2001, at http://www.asam.org/advocacy/find-a-policy-statement/view-policy-statement/public-policy-state ments/2011/12/15/the-definition-of-addiction.

22. Ernest Kurtz, *Alcoholics Anonymous and the Disease Concept of Addiction*, available as an e-book at http://ebookbrowse.com/ernie-kurtz-aa-the-disease-concept-of-alcoholism-pdf -d168865618.

23. Personal experiences of Sally Satel in her work as a psychiatrist over the years in methadone clinics.

24. Thomas C. Schelling, "The Intimate Contest for Self-command," *Public Interest* 60 (1980): 94–118; Robert Fagles, trans., *The Odyssey of Homer* (New York: Penguin Books, 1997), 272–273; "Historical Perspectives: Opium, Morphine, and Heroine," Wired into Recovery, http:// wiredintorecovery.org/articles/entry/8932/historical-perspectives-opium-morphine-and-heroin/; Southwest Associates, "How Interventions Work," http://www.southworthassociates.net/interven tions/how-interventions-work.

25. Edward J. Khantzian and Mark J. Albanese, *Understanding Addiction as Self Medication: Finding Hope Behind the Pain* (Lanham, MD: Rowman and Littlefield, 2008); Caroline Knapp, *Drinking: A Love Story* (New York: Dial Press, 1997), 267. Knapp died at age forty-two from lung cancer after being sober for several years.

26. Jerry Stahl, *Permanent Midnight* (Los Angeles: Process, 2005), 6, 3.

27. *Cracked Not Broken*, DVD, directed by Paul Perrier (HBO, 2007).

28. Harold Kalant, "What Neurobiology Cannot Tell Us About Addiction," *Addiction* 105, no. 5 (2010): 780–789; Nick Heather, "A Conceptual Framework for Explaining Drug Addiction," *Journal of Psychopharmacology* 12, no. 1 (1998): 3–7. British actor Russell Brand, a former heroin addict, reflected on the death of his friend Amy Winehouse, "The priority of any addict is to anesthetize the pain of living to ease the passage of the day with some purchased relief." "Russell Brand Pens Touching Tribute to Amy Winehouse," *US Weekly*, July 24, 2011, http://www.usmagazine .com/entertainment/news/russell-brand-pens-touching-tribute-to-amy-winehouse-2011247. Similarly, Pete Hamill writes in his memoir *A Drinking Life*, "The culture of drink endures because it offers so many rewards: confidence for the shy, clarity for the uncertain, solace to the wounded and lonely." Hamill, *A Drinking Life* (New York: Back Bay Books, 1995), 1. Marc Lewis recalls, "My own recovery was partly based on how terrible I eventually came to feel when I was on drugs. Whatever rewards that being high had once had were no longer available. Being so immersed in constant drug taking became associated with so many awful feelings that it was a far worse experience than anything sobriety had to offer." Walter Armstrong, "Interview with an Addicted Brain," *The Fix*, May 23, 2012, http://www.thefix.com/content/interview-Marc-Lewis -addicted-brain8090?page=all.

29. *Powell v. Texas*, 392 U.S. 14 (1968), http://bulk.resource.org/courts.gov/c/US/392/392.US .514.405.html.

30. Ibid.

31. Ibid.

32. Stephen T. Higgins, Kenneth Silverman, and Sarah H. Heil, eds. *Contingency Management in Substance Abuse Treatment* (New York: Guilford Press, 2008). A meta-analysis found that rewards produced a 61 percent success rate (for all drugs combined) in the experimental condition versus a 39 percent success rate in the treatment-as-usual condition. Michael Prendergast et al., "Contingency Management for Treatment of Substance Use Disorders: A Meta-analysis," *Addiction* 101, no. 11 (2006): 1546–1560; and Kevin G. Volpp et al., "A Randomized, Controlled Trial of Financial Incentives for Smoking Cessation," *New England Journal of Medicine* 360 (2009): 699–709.

33. Some evidence suggests that fMRI reactivity to substance-related cues has predictive value regarding relapse. In one study with cocaine addicts, it was the observation of greater activation of the limbic areas before the subject's report of subjective craving responses—and not the subjective reports of craving—that predicted which patients would relapse during a ten-week treatment trial. This is interesting and of potential clinical value, especially if risk of relapse can't be gauged in a less expensive way. Thomas R. Kosten et al., "Cue-Induced Brain Activity Changes and Relapse in Cocaine-Dependent Patients," *Neuropsychopharmacology* 31 (2006): 644–650. Functional imaging studies can help identify patients who display strong physiological cue reactivity and who are thus at risk of suffering a relapse when they are confronted with alcohol-associated cues. See Andreas Heinz et al., "Brain Activation Elicited by Affectively Positive Stimuli Is Associated with a Lower Risk of Relapse in Detoxified Alcoholic Subjects," *Alcoholism: Clinical and Experimental Research* 31, no. 7 (2007): 1138–1147; and Amy C. Janes et al., "Brain Reactivity to Smoking Cues Prior to Smoking Cessation Predicts Ability to Maintain Tobacco Abstinence," *Biological Psychiatry* 67 (2010): 722–729. See also Daniel Shapiro, cited in Sally Satel and Frederick Goodwin, "Is Addiction a Brain Disease?," Ethics and Public Policy, 1997, www.eppc.org/docLib/20030420_DrugAddictionBrainDisease.pdf.

34. Hedy Kober et al., "Prefrontostriatal Pathway Underlies Cognitive Regulation of Craving," *Trends in Cognitive Neuroscience* 15, no. 3 (2011): 132–139. See also Cecilia Westbrook et al., "Mindful Attention Reduces Neural and Self-Reported Cue-Induced Craving in Smokers," *Social Cognition and Affective Neuroscience*, 2011, http://scan.oxfordjournals.org/content/early/2011/11/22/scan.nsr076.full; Angela Hawken, "Behavioral Triage: A New Model for Identifying and Treating Substance-Abusing Offenders," *Journal of Drug Policy Analysis* 3, no. 1 (February 2010), doi: 10.2202/1941-2851.1014.

35. Nora D. Volkow et al., "Cognitive Control of Drug Craving Inhibits Brain Reward Regions in Cocaine Abusers," *Neuroimage* 49 (2010): 2536–2543.

36. Robert L. DuPont et al., "Setting the Standard for Recovery: Physicians' Health Programs," *Journal of Substance Abuse Treatment* 36, no. 2 (2009): 159–171, 165. See also Robert L. DuPont et al., "How Are Addicted Physicians Treated? A National Survey of Physician Health Programs," *Journal of Substance Abuse Treatment* 37, no. 1 (2009): 1–7.

37. Maxine Stitzer and Nancy Petry, "Contingency Management for Treatment of Substance Abuse," *Annual Review of Clinical Psychology* 2 (2006): 411–434. Note that one study found that cocaine-dependent subjects were less risky on a decision-making gambling task (at the same level as control subjects) when they could win real money than when only hypothetical gains and losses were at stake; see Nehal P. Vadhan et al., "Decision-making in Long-Term Cocaine Users: Effects of a Cash Monetary Contingency on Gambling Task Performance," *Drug and Alcohol Dependence* 102, no. 103 (2009): 95–101.

38. Beginning in 1989, drug courts were established in Florida, and there are now several thousand of them across the country. They were designed to give nonviolent addicts the option of pleading guilty and entering a treatment program, closely overseen by the drug court judge. Nonviolent defendants are typically required to plead guilty or no contest, but drug court offers them

the chance to have their criminal record expunged if they complete a program in which they are closely monitored by a judge for at least a year. If patient-offenders fail a random urine screen or violate program rules (including attendance at treatment), the judge reliably administers swift, certain, but not severe sanctions, such as community service or a night in jail. The sanctions are also fairly administered and transparent: All individuals are treated the same for the same behavior, and they know exactly what will happen in case of an infraction. Drug court dropout rates are significantly lower than dropout rates of patients in standard treatment. This is important because it is well established that good treatment outcomes, such as a decline in drug use, are tightly linked to the length of time a person spends in treatment regardless of whether that treatment was undertaken as a condition of a court or freely chosen by the patient. Furthermore, compared with probationers undergoing standard probationary conditions, criminal recidivism among drug court participants is significantly reduced. There is considerable variability from court to court in laxity, funding, staffing, size of program, and when sanctions are triggered. For a description of drug courts, see Shannon M. Carey, Michael W. Finigan, and Kimberly Pukstas, *Exploring the Key Components of Drug Courts: A Comparative Study of 18 Adult Drug Courts on Practices, Outcomes, and Costs* (Portland, OR: NPC Research, 2008), electronic copy available at http://www.ncjrs.gov/pdffiles1 /nij/grants/223853.pdf; National Drug Court Research Center, "How Many Drug Courts Are There?," updated December 31, 2011, http://www.ndcrc.org/node/348; and Celinda Franco, "Drug Courts: Background, Effectiveness, and Policy Issues for Congress," Congressional Research Service, October 12, 2010, http://www.fas.org/sgp/crs/misc/R41448.pdf. On Project HOPE, see Angela Hawken and Mark Kleiman, "Managing Drug Involved Probationers with Swift and Certain Sanctions: Evaluating Hawaii's HOPE," December 2, 2009, http://www.nij.gov/topics/corrections /community/drug-offenders/hawaii-hope.htm.

39. Angela Hawken, "Behavioral Triage." Note that patients were referred for treatment if they tested positive three or more times over the one-year follow-up period. Angela Hawken, School of Public Policy at Pepperdine University, personal communication with authors, February 16, 2012. Notably, HOPE is mandatory for those selected; it's not optional, but it's also not universal. Unlike other programs, participants are selected for high risk of failure without the program rather than high probability of success on it.

40. Angela Hawken and Mark Kleiman, "Managing Drug Involved Probationers with Swift and Certain Sanctions: Evaluating Hawaii's HOPE," December 2, 2009, http://www.nij.gov/topics/cor rections/community/drug-offenders/hawaii-hope.htm. On the effects of methamphetamines, see Ari D. Kalechstein, Thomas F. Newton, and Michael Green, "Methamphetamine Dependence Is Associated with Neurocognitive Impairment in the Initial Phases of Abstinence," *Journal of Neuropsychiatry and Clinical Neuroscience* 15 (2003): 215–220; Thomas E. Nordahl, Ruth Salo, and Martin Leamon, "Neuropsychological Effects of Chronic Methamphetamine Use on Neurotransmitters and Cognition: A Review," *Journal of Neuropsychiatry and Clinical Neuroscience* 15 (2003): 317–25; Patricia A. Woicik et al., "The Neuropsychology of Cocaine Addiction: Recent Cocaine Use Masks Impairment," *Neuropsychopharmacology* 34, no. 5 (2009): 1112–1122; and Mark S. Gold et al., "Methamphetamine- and Trauma-Induced Brain Injuries: Comparative Cellular and Molecular Neurobiological Substrates," *Biological Psychiatry* 66, no. 2 (2009): 118–127; Carl L. Hart et al., "Cognitive Functioning Impaired in Methamphetamine Users? A Critical Review," *Neuropsychopharmacology* 37 (2012): 586–608 (note that the paper by Hart et al. reports on recreational rather than chronic users of methamphetamine).

41. For example, South Dakota sought to reduce the number of repeat drunk drivers. It required individuals arrested for, or convicted of, alcohol-related offenses to submit to breathalyzer tests twice per day or wear a continuous alcohol-monitoring bracelet. Those testing positive were subject to swift, certain, and moderate sanctions and were arrested on the spot, tried that day, and sentenced to a night in jail. From 2005 to 2010, program participants were ordered to take approximately 3.7 million breathalyzer tests, and the pass rate exceeded 99 percent. See Beau Kilmer

et al., "Efficacy of Frequent Monitoring with Swift, Certain, and Modest Sanctions for Violations: Insights from South Dakota's 24/7 Sobriety Project," *American Journal of Public Health* 103, no. 1 (2013): e37-e43. Not only are incentives more effective than treatment as usual, but alone they can sometimes work as well as, or even better than, incentives plus treatment. Using incentives can also result in significant savings for taxpayers. Adele Harrell, Shannon Cavanagh, and John Roma, *Findings from the Evaluation of the D.C. Superior Court Drug Intervention Program* (Washington, DC: Urban Institute, 1999), https://www.ncjrs.gov/pdffiles1/nij/grants/181894.pdf. Finally, contrary to conventional wisdom, addicts don't have to want to change their lives in order for a treatment program to succeed. If they stay in it long enough—and the leverage afforded by incentives increases this likelihood—most eventually absorb the values of the program as they appreciate the benefits of drug-free living. Indeed, addicts who are legally pressured into treatment may outperform voluntary patients because they are likely to stay in treatment longer and are more likely to graduate. Sally L. Satel, *Drug Treatment: The Case for Coercion* (Washington, DC: AEI Press, 1999); Judge Steven Alm, personal communication with authors, June 30, 2009. A number of drug courts, a type of contingency-management, criminal justice diversion program for nonviolent offenders, have also been hampered by staff resistance to penalties for positive urine tests. Linda L. Chezem, J. D., Adjunct Professor, Department of Medicine at Indiana University School of Medicine, personal communication with authors, February 10, 2009. On NIDA's refusal to review Project HOPE, see Mark Kleiman, "How NIDA Puts the Dope Back into Dopamine," The Reality-Based Community, July 19, 2012, www.samefacts.com/2012/07/drug-policy/how-nida-puts-the-dope-back-into-dopamine. The same logic—addicts' brain changes make them insensitive to incentives—was also the destructive logic employed by the opponents of Proposition 36 in California, a 2001 referendum on the state's jail diversion program for nonviolent drug offenders. They prevailed, and within a few years treatment-program staff began clamoring for permission to use modest penalties and incentives—without them, the staff had little leverage.

42. "The absence of medications for stimulant addiction is probably at the core of our inability to get a handle on this issue in this country, and I have declared the development of anti-stimulant addiction medications in my institution as a top priority." Alan I. Leshner, "Treatment: Effects on the Brain and Body," National Methamphetamine Drug Conference, May 29–30, 1997, https://www.ncjrs.gov/ondcppubs/publications/drugfact/methconf/plenary2.html. Similarly, Leshner writes, "Our ultimate goal is to apply the knowledge we gain from our brain imaging studies to the development of better targeted, more effective treatments for drug abuse and addiction." "Director's Column: NIDA's Brain Imaging Studies Serve as Powerful Tools to Improve Drug Abuse Treatment," *NIDA Notes* 11, no. 5 (1996), http://archives.drugabuse.gov/NIDA_Notes/NNVol11N5/DirRepVol11N5.html. Interlandi, "What Addicts Need."

43. Melanie Greenberg, "Could Neuroscience Have Helped Amy Winehouse?," *Psychology Today*, July 24, 2011, http://www.psychologytoday.com/blog/the-mindful-self-express/201107/could-neuroscience-have-helped-amy-winehouse. Science writer Sharon Begley notes the disproportionate media attention to medical cures for addiction: "The media laud the smallest steps toward a cocaine vaccine, when a study shows—yawn—that rewarding abstinence can get addicts off meth, that couples counseling can treat alcoholism, or that cognitive-behavior therapy can break the grip of coke addiction . . . silence." Sharon Begley, "Forget the Cocaine Vaccine," *Newsweek*, March 4, 2010, http://www.thedailybeast.com/newsweek/2010/03/04/forget-the-cocaine-vaccine.html. David M. Eagleman, Mark A. Correro, and Jyotpal Sing, "Why Neuroscience Matters for Rational Drug Policy," *Minnesota Journal of Law, Science and Technology* 11, no. 1 (2010): 7–26.

44. National Institute on Drug Abuse, *Drugs, Brains and Behavior: The Science of Addiction* (Bethesda, MD: National Institutes of Health, 2007): 1, http://www.drugabuse.gov/sites/default/files/sciofaddiction.pdf. But in truth, not a single recent brain-based discovery has yielded a game-changing treatment advance. The most used medications first appeared before the 1980s. On methadone, see David F. Musto, *The American Disease: Origins of Narcotics Control* (Oxford:

Oxford University Press, 1999), 237–253. The prevailing understanding at the time was that once a heroin addict, always an addict. Vincent Dole and Marie Nyswander, the husband-and-wife physician team who pioneered methadone treatment, considered heroin addiction analogous to diabetes and methadone maintenance analogous to insulin for a diabetic. On continued use of drugs by people on methadone maintenance, see Edward J. Cone, "Oral Fluid Results Compared to Self-Report of Recent Cocaine and Heroin Use by Methadone Maintenance Patients," *Forensic Science International* 215 (2012): 88–91. For a good summary of cocaine immunotherapy, see Thomas Kosten, "Shooting Down Addiction," *Scientist Daily*, June 1, 2011, http://the-scientist .com/2011/06/01/shooting-down-addiction/. The goal of the cocaine vaccine is to stimulate the body to produce antibodies to cocaine. These combine with cocaine molecules if an individual uses the drug, forming a complex that is too bulky to penetrate the brain's vasculature into the brain tissue. Presumably, a vaccinated addict will soon lose interest in cocaine because the drug would have no effect. The idea of using vaccines to fight drug addiction traces its origins back nearly forty years, to tests in the early 1970s on mice and monkeys. Jerome Jaffe, personal communication with authors, January 15, 2011. Long-acting versions of naltrexone in the form of a monthly injection or pellet implanted under the skin have been developed to circumvent the problem of "forgetting" to take the medication every day. Methadone also has opiate-blocking properties because it occupies relevant opiate receptors. A medication called buprenorphine is a partial opiate blocker. M. Srisurapanont and N. Jarusuraisin, "Opioid Antagonists for Alcohol Dependence," *Co-chrane Database of Systematic Reviews* 25, no. 1 (2005): CD001867.

45. George F. Koob, G. Kenneth Lloyd, and Barbara J. Mason, "The Development of Pharmaco-therapies for Drug Addiction: A Rosetta Stone Approach," *Nature Reviews Drug Discovery* 8, no. 6 (2009): 500–515. The mechanisms of action of naltrexone in alcoholics may be manifold. First, naltrexone can reduce craving, which is the urge or desire to drink. Second, naltrexone helps patients remain abstinent. Third, naltrexone may interfere with the tendency to want to drink more if a recovering patient has a drink, presumably by interrupting endorphin-mediated pleasure. Silvia Minozzi et al., "Oral Naltrexone Maintenance Treatment for Opiate Dependence," *Cochrane Database of Systematic Reviews* 16, no. 2 (2011): CD001333. Acamprosate is another medication that reduces craving for alcohol, but the mechanism of action is unclear. "Acamprosate: A New Medication for Alcohol Use Disorders," *Substance Abuse Treatment Advisory* 4, no. 1 (2005), http://kap.samhsa.gov/products/manuals/advisory/pdfs/Acamprosate-Advisory.pdf. The big challenge for medication development is how to disrupt the dopamine-rich mesolimbic system that mediates craving without affecting enjoyment of natural rewards such as sex and food, which is dependent on an intact reward mechanism. For another interesting side effect of dopamine enhancement, see Leann M. Dodd et al., "Pathological Gambling Caused by Drugs Used to Treat Parkinson Disease," *Archives of Neurology* 62, no. 9 (2005): 1377–1381.

46. Alan Leshner's outlook, described in Peggy Orenstein, "Staying Clean," *New York Times Magazine*, February 10, 2002; Leshner, "Addiction Is a Brain Disease," 46 and 45; Nora Volkow, "It's Time for Addiction Science to Supersede Stigma," *ScienceNews*, October 24, 2008.

47. "As we relate to the public and policy makers," said Glen Hanson, a former NIDA official, "[we must] help them understand that this is a disease process . . . rather than an issue of decision-making." Glen R. Hanson, "How Casual Drug Use Leads to Addiction: The 'Oops' Phenomenon," *Atlanta Inquirer* 41, no. 5 (2002): 4.

48. Susan Cheever, "Drunkenfreude," Proof, *New York Times*, December 15, 2008, http://proof .blogs.nytimes.com/2008/12/15/drunkenfreude/.

49. A 2006 HBO, USA Today, and Gallup poll found that 66 percent of family members said that addiction is both a psychic and a mental disease (8 percent said that it is only a psychic one). Twenty-four percent said that it is not a disease. The same poll also found that 55 percent of family members ranked "lack of will power" as the top factor in addiction (with 60 percent of family members saying that it is the top factor for drug addiction, specifically). Only about one-sixth, or 16

percent, said that it was not a factor at all; it was thus the factor least likely to be dismissed as wholly irrelevant. Half thought that depression or anxiety played the most important roles in contributing to addiction; see "USA Today Poll," May 2006, http://www.hbo.com/addiction/understanding addiction/17 usa today poll.html. The Substance Abuse Mental Health Services Administration (HHS) 2008 Caravan Study found that 66 percent of eighteen- to twenty-four-year-olds think that people could stop drugs if they had enough willpower, but the general population says 38 percent; see "National Poll Reveals Public Attitudes on Substance Abuse, Treatment and the Prospects of Recovery," Substance Abuse and Mental Health Services Administration (SAMHSA), last modified December 4, 2006, http://www.samhsa.gov/attitudes/. A 2005 Peter D. Hart study on alcoholism only found that 63 percent consider it a moral weakness and 37 percent a disease, but the majority favor treatment; see Alan Rivlin, "Views on Alcoholism & Treatment," September 29, 2005, http://www.facesandvoicesofrecovery.org/pdf/2005-09-29_rivlin_presentation.pdf. A 2008 Hazelden poll, National Study of Public Attitudes Towards Addiction, found that 78 percent "agree" (44 percent) and "strongly agree" (34 percent) that addiction is a disease, and 56 percent say that there should be no prison for a first offense; see www.hazelden.org/web/public/document/2008publicsurvey.pdf. A 2009 Open Society Institute poll found that 75 percent think that addiction is a disease; www.facesandvoicesofrecovery.org/pdf/OSI_LakeResearch_2009.pdf. A 2007 Robert Wood Johnson Foundation survey found that 47 percent see addiction as a "form of illness." Nonwhites were significantly more likely to see it as a "personal weakness"; about 13 percent of all respondents see it as both a weakness and a disease. Two-thirds think that recovery is not possible without professional help or AA-like assistance. See "What Does America Think About Addiction Prevention and Treatment?," *RWJF Research Highlight* 24 (2007), https://folio.iupui.edu/ . . . /559/Research%20Highlight %2024[3].pdf. The Indiana University study is reported in Bernice A. Pescosolido et al., "'A Disease Like Any Other'? A Decade of Change in Public Reactions to Schizophrenia, Depression, and Alcohol Dependence," *American Journal of Psychiatry* 167, no. 11 (2010): 1321–1330.

50. Gau Schomerus et al., "Evolution of Public Attitudes About Mental Illness: A Systematic Review and Meta-analysis," *Acta Psychiatrica Scandinavica* 125, no. 6 (2012): 440–452; Daniel C. K. Lam and Paul Salkovskis, "An Experimental Investigation of the Impact of Biological and Psychological Causal Explanations on Anxious and Depressed Patients' Perception of a Person with Panic Disorder," *Behaviour Research and Therapy* 45 (2006): 405–411; John Read and Niki Harré, "The Role of Biological and Genetic Causal Beliefs in the Stigmatisation of 'Mental Patients,'" *Journal of Mental Health* 10 (2001): 223–235; John Read and Alan Law, "The Relationship of Causal Beliefs and Contact with Users of Mental Health Services to Attitudes to the 'Mentally Ill,'" *International Journal of Social Psychiatry* 45 (1999): 216–229; Danny C. K. Lam and Paul M. Salkovskis, "An Experimental Investigation of the Impact of Biological and Psychological Causal Explanations on Anxious and Depressed Patients' Perception of a Person with Panic Disorder," *Behavior Research and Therapy* 45, no. 2 (2007): 405–411; Sheila Mehta and Amerigo Farina, "Is Being 'Sick' Really Better? Effect of the Disease View of Mental Disorder on Stigma," *Journal of Social and Clinical Psychology* 16, no. 4 (1997): 405–419; Ian Walker and John Read, "The Differential Effectiveness of Psychosocial and Biogenetic Causal Explanations in Reducing Negative Attitudes Toward 'Mental Illness,'" *Psychiatry* 65, no. 4 (2002): 313–325; Brett J. Deacon and Grayson L. Baird, "The Chemical Imbalance Explanation of Depression: Reducing Blame at What Cost?," *Journal of Social and Clinical Psychology* 28, no. 4 (2009): 415–435; Matthias C. Angermeyer and Herbert Matschinger, "Labeling—Stereotype—Discrimination: An Investigation of the Stigma Process," *Social Psychiatry and Psychiatric Epidemiology* 40 (2005): 391–395; and Nick Haslam, "Genetic Essentialism, Neuroessentialism, and Stigma: Commentary on Dar-Nimrod and Heine (2011)," *Psychological Bulletin* 137 (2011): 819–824.

51. Leshner, "Addiction Is a Brain Disease," 47.

52. Under the official criteria of the *Diagnostic and Statistical Manual* (*DSM*) for "substance-related disorder" called dependence (which is effectively interchangeable with the term "addiction"),

any three of the following nine symptoms are sufficient for a "dependence" diagnosis: (1) substance taken in larger amounts or for longer period than intended, (2) persistent desire or one or more unsuccessful efforts to cut down or control substance use, (3) a great deal of time spent obtaining or consuming the substance or recovering from its effects, (4) frequent intoxication or withdrawal when expected to fulfill major role obligations, (5) important social, occupational, or recreational activities given up because of use, (6) continued use despite recurrent problems related to use, (7) tolerance, (8) withdrawal, and (9) use in order to avoid withdrawal. *Diagnostic and Statistical Manual of Mental Disorders: DSM-IV-TR,* 4th ed.; text revision (Washington, DC: American Psychiatric Publishing, 2000), 193.

Chapter 4

1. Angela Saini, "How India's Neurocops Used Brain Scans to Convict Murderers," *Wired* 6 (2009), http://www.wired.co.uk/wired-magazine/archive/2009/05/features/guilty.aspx?page=all. India uses BEOS and brain fingerprinting, which we describe presently in the text, as well as the polygraph and truth serum. Brain fingerprinting came first; BEOS is an offshoot. Sharma went through both brain fingerprintng and BEOS. The BEOS technicians claim to use more brain information than P300, but they have kept their actual method proprietary. Hank T. Greely, personal communication, November 29, 2012.

2. Saini, "How India's Neurocops Used Brain Scans"; The judge proclaimed that BEOS "clearly indicated her involvement in the murder of Udit." *State of Maharashtra v. Aditi Baldev Sharma and Pravin Premswarup Khandelwal,* Sessions Case No. 508/07 (2008): 61, http://lawandbiosciences .files.wordpress.com/2008/12/beosruling2.pdf; Hank T. Greely, personal communication November 29, 2012.

3. Rosenfeld is quoted and statistics on BEOS use are given in Saini, "How India's Neurocops Used Brain Scans to Convict Murderers." In 2010, the Indian Supreme Court ruled that the procedure, if it was performed without the suspect's consent, violated his or her right to protection against self-incrimination. The court, however, still permitted the test's use if the accused consented and it was corroborated by other evidence. Dhananjay Mahapatra, "No Narcoanalytics Test Without Consent, Says SC," *Times of India,* May 5, 2010, http://articles.timesofindia.india times.com/2010-05-05/india/28319716_1_arushi-murder-case-nithari-killings-apex-court. For "neurocops," see Saini, "How India's Neurocops Used Brain Scans"; for "thought police," see Helen Pearson, "Lure of Lie Detectors Spooks Ethicists," *Nature* 441 (2006): 918–919; for "brain-jacked," see John Naish, "Can a Machine Read Your Mind?," *Times* (London), February 28, 2009. "Mukundan will not disclose the inner workings of his brain-imaging software. His decision not to publish his research or subject his ideas to peer review has prevented others from verifying his results. But he does not care, he says, because he would rather his peers condemned him than he lost control of his invention before it is patented." Saini, "How India's Neurocops Used Brain Scans."

4. The review of BEOS analysis was initiated in May 2007. See M. Raghava, "Stop Using Brain Mapping for Investigation and as Evidence," *Hindu,* September 6, 2008, http://www.hindu.com /2008/09/06/stories/2008090655050100.htm. The conclusion of the Indian National Institute of Mental Health and Neurosciences was released in 2008. Sharma had appealed her conviction, and in April 2009, her sentence was suspended and bail was granted pending disposal of the appeal. This decision was based on questions about her possession of arsenic. See the decision of the High Court of Judicature at Bombay for Criminal Application No. 1294 of 2008, http://lawandbiosciences .files.wordpress.com/2009/04/iditis-bail-order1.pdf.

5. Saini, "How India's Neurocops Used Brain Scans." Sharma was released because the evidence that she possessed the arsenic-laced prasad was not compelling, and indeed, "the possibil-

ity of plantation cannot not be ruled out" [*sic*]. See Emily Murphy, "Update on Indian BEOS Case: Accused Released on Bail," April 2, 2009, http://lawandbiosciences.wordpress.com/2009/04 /02/update-on-indian-beos-case-accused-released-on-bail/ (with reprint of her bail release document).

6. Concise general overviews: Daniel D. Langleben and Jane C. Moriarty, "Using Brain Imaging for Lie Detection: Where Science, Law, and Policy Collide," *Psychology, Public Policy, and Law* (forthcoming). For a report on the state of the art and on being cautious, see Brandon Keim, "Brain Scanner Can Tell What You're Looking At," *Wired*, March 5, 2008, http://www.wired.com /science/discoveries/news/2008/03/mri_vision. Steve Silberman, in his article "Don't Even Think About Lying," *Wired*, January 2006, http://www.wired.com/wired/archive/14.01/lying.html, proclaimed that fMRI is "poised to transform the security industry, the judicial system, and our fundamental notions of privacy." Such talk has made brain-scan lie detection sound as solid as DNA evidence, which it most definitely is not. See also "Neuroscientist Uses Brain Scan to See Lies Form," National Public Radio, October 30, 2007, http://www.npr.org/templates/transcript/tran script.php?storyId=15744871; and *Newsweek's* 2008 announcement that "mind reading has begun" in Sharon Begley, "Mind Reading Is Now Possible," *Newsweek*, January 12, 2008, http://www .newsweek.com/id/91688/page/2. The No Lie MRI statement is at http://noliemri.com/customers /Overview.htm.

7. P. R. Wolpe, K. R. Foster, and D. D. Langleben, "Emerging Neurotechnologies for Lie-Detection: Promises and Perils," *American Journal of Bioethics* 5 (2005): 39–49.

8. Charles V. Ford, *Lies! Lies! Lies! The Psychology of Deceit* (Washington, DC: American Psychiatric Publishing, 1999); Paul V. Trovillo, "A History of Lie Detection," *American Journal of Police Science* 29, no. 6 (1939): 848–881. Freud said, "No mortal can keep a secret. If his lips are silent, he chatters with his fingertips; betrayal oozes out of him at every pore"; quoted in *The Freud Reader*, ed. Peter Gay (New York: W. W. Norton, 1995), 215. On the belief that one can spot a liar, see Global Deception Research Team, "A World of Lies," *Journal of Cross-Cultural Psychology* 37, no. 1 (2006): 60–74. On inability to detect lies, see Paul Ekman, Maureen O'Sullivan, and Mark G. Frank, "A Few Can Catch a Liar," *Psychological Science* 10, no. 3 (1999): 263–266; and Aldert Vrij et al., "Detecting Lies in Young Children, Adolescents, and Adults," *Applied Cognitive Psychology* 20, no. 9 (2006): 1225–1237.

9. The rate of lying was found by researchers who asked subjects aged eighteen to seventy-one to keep a diary of all the falsehoods they told over the course of a week; see Bella M. DePaulo, "The Many Faces of Lies," in *The Social Psychology of Good and Evil*, ed. A. G. Miller (New York: Guilford Press, 2004), 303–326, available on page 4 at http://smg.media.mit.edu/library /DePaulo.ManyFacesOfLies.pdf. On English words for deception, see Robin Marantz Henig, "Looking for the Lie," *New York Times*, February 5, 2006, www.nytimes.com/2006/02/05/maga zine/05lying.html. On deception across cultures, see Sean A. Spence et al., "A Cognitive Neurobiological Account of Deception: Evidence from Functioning Neuroimaging," *Philosophical Transactions of the Royal Society of London B* 359 (2004): 1755–1762, 1756.

10. Robert Trivers, *The Folly of Fools: The Logic of Deceit and Self-Deception in Human Life* (New York: Basic Books, 2011); Alison Gopnik, *The Philosophical Baby: What Children's Minds Tell Us About Truth, Love, and the Meaning of Life* (New York: Picador, 2010), 54–61. Some nonhuman primates may possess a nascent capacity for instilling false beliefs in the minds of others, but, as far as we know, only humans are preoccupied with trying to figure out whether they have been lied to. See Robert W. Byrne, "Tracing the Evolutionary Path of Cognition: Tactical Deception in Primates," in *The Social Brain: Evolution and Pathology*, ed. M. Brüne, H. Ribbert, and W. Schiefenhövel (Hoboken, NJ: John Wiley, 2003). Individuals with autism, by contrast, who lack a fully developed theory of mind, are unconvincing liars and are generally poor at recognizing deception in others. Simon Baron-Cohen, "Out of Sight or Out of Mind: Another Look at Deception in Autism," *Journal of Child Psychology and Psychiatry* 33, no. 7 (1992): 1141–1155.

11. David T. Lykken, *A Tremor in the Blood: Use and Abuses of the Lie Detector* (New York: Basic Books, 1998); Anne M. Bartol and Curt R. Bartol, *Introduction to Forensic Psychology: Research and Application* (Thousand Oaks, CA: Sage, 2012), 101.

12. Responses to an examiner's questions were etched by a stylus onto paper fitted to a revolving drum. In 1938, the Gillette Company hired Marston to prove that freshly shaven men were telling the truth when they said that the Gillette blade was better than a competitor's. Marston claimed that he found this to be so, but his data were later proved fraudulent. If only Gillette had used that Golden Lasso on Marston. Les Daniels, *Wonder Woman: The Complete History* (San Francisco: Chronicle Books, 2004), 16; Mark Constanzo and Daniel Krauss, *Forensic and Legal Psychology: Psychological Science Applied to Law* (New York: Worth, 2012), 55.

13. Ken Alder, *The Lie Detectors: The History of an American Obsession* (New York: Free Press, 2007), xi. Some 5,000 to 10,000 polygraph operators were testing 2 million Americans each year during the 1980s (ibid., xiv). According to Alder, "The lie detector had become America's mechanical conscience" (ibid., xiv). "What began as a way to confirm honesty in precinct stations, office towers, and government agencies became a way to test the credibility of Hollywood movies and Madison Avenue advertising" (ibid., xiii–xiv). See also *Hearings on the Use of Polygraphs as "Lie Detectors" by the Federal Government Before the House Comm. on Government Operations*, 88th Cong., 2d Sess. (1964); H.R. Rep. No. 198, 89th Cong., 1st Sess., 13 (1965). For an overview on *Daubert*, see Martin C. Calhoun, "Scientific Evidence in Court: Daubert or Frye, 15 Years Later," Washington Legal Foundation, *Legal Backgrounder*, vol. 23, no. 37, August 22, 2008, http://www.wlf.org/upload/08-22-08calhoun.pdf. *Frye v. United States*, 293 F. 1013, 1014 (D.C. Cir. 1923) (holding that scientific evidence must use methods generally accepted in the relevant scientific community), http://law.jrank.org/pages/12871/Frye-v-United-States.html; *Daubert v. Merrell-Dow Pharmaceuticals*, 509 U.S. 579 (1993). Further specifications regarding the admissibility of expert witnesses were added by the Supreme Court in *Kumho Tire Co. v. Carmichael*, 526 U.S. 137 (1999), and *General Electric Co. v. Joiner*, 522 U.S. 136 (1997).

14. See Employee Polygraph Protection Act, 29 U.S.C. § 22, (1988), http://finduslaw.com/employee_polygraph_protection_epp_29_u_s_code_chapter_22#6; Joan Biskupic, "Justices Allow Bans of Polygraph," *Washington Post*, April 1, 1998. The polygraph is also used in criminal investigations, even though the results can't be used in court, and as a routine part of probation for sex offenders; http://www.polygraph.org/section/resources/frequently-asked-questions. Roughly 1.6 million tests have been performed yearly during the first years of the twenty-first century, up about 50 percent from a decade ago, the American Polygraph Association told the *Wall Street Journal*. Laurie P. Cohen, "The Polygraph Paradox," *Wall Street Journal*, March 22, 2008, http://online.wsj.com/article/SB120612863077155601.html (mostly used by the feds).

15. National Research Council, Division of Behavioral and Social Sciences and Education, *The Polygraph and Lie Detection* (Washington, DC: National Academies Press, 2003), esp. 3.

16. Allan S. Brett, Michael Phillips, and John F. Beary, "The Predictive Power of the Polygraph: Can the 'Lie Detector' Really Detect Liars?," *Lancet* 327, no. 8480 (1986): 544–547. In an exhaustive review of the evidence in 2003, the National Research Council found that although polygraph accuracy was "better than chance," there was "little basis for the expectation [of] extremely high accuracy." *Polygraph and Lie Detection*, 212. On Ames and Lee, see "The C.I.A. Security Blanket," *New York Times*, September 17, 1995; and Vernon Loeb and Walter Pincus, "FBI Misled Wen Ho Lee into Believing He Failed Polygraph," *Washington Post*, January 8, 2000.

17. Other ways to detect recognition are through temperature patterns around the subjects' eyes, reflecting facial blood flow; or cortical blood flow underneath the scalp, registered by sensors using infrared light. See G. Ben-Shakar and E. Elaad, "The Validity of Psychophysiological Detection of Information with the Guilty Knowledge Test: A Meta-analytic Review," *Journal of Applied Psychology* 88, no. 1 (2003): 131–151.

18. L. A. Farwell and E. Donchin, "The Truth Will Out: Interrogative Polygraphy ('Lie Detection') with Event-Related Potentials," *Psychophysiology* 28 (1991): 531–547; L. A. Farwell et al., "Optimal Digital Filters for Long Latency Components of the Event-Related Brain Potential," *Psychophysiology* 30 (1993): 306–315; L. A. Farwell, "Method for Electroencephalographic Information Detection," *U.S. Patent*, no. 5 (1995): 467, 777; Lawrence A. Farwell and Sharon S. Smith, "Using Brain MERMER Testing to Detect Knowledge Despite Efforts to Conceal," *Journal of Forensic Sciences* 46, no. 1 (2001): 135–143. The MERMER comprises a P300 response, occurring 300 to 800 milliseconds after the stimulus, and additional patterns occurring more than 800 milliseconds after the stimulus, providing even more accurate results; see Farwell and Smith, "Using Brain MERMER Testing."

19. For a detailed critique of Farwell's claims, see J. Peter Rosenfeld, "'Brain Fingerprinting': A Critical Analysis," *Scientific Review of Mental Health Practice* 4, no. 1 (2005): 20–37. For a rebuttal, see Lawrence A. Farwell, "Brain Fingerprinting: Corrections to Rosenfeld," *Scientific Review of Mental Health Practice* 8, no. 2 (2011): 56–68. In October 2001, a General Accounting Office report on brain fingerprinting found that both the CIA and the FBI believed that the technology had "limited applicability and usefulness." Each cited Farwell's reluctance to provide algorithmic information about how his test works on the grounds that his technique was proprietary. Becky McCall, "Brain Fingerprints Under Scrutiny," *BBC News*, February 17, 2004, http://news.bbc.co.uk/2/hi/science/nature/3495433.stm. On Slaughter's execution, see Doug Russell, "Family's Nightmare Ends as Murderer Executed," *McAlester News Democrat*, March 16, 2005. For details on the Oklahoma Court of Appeals' decision to deny the evidentiary hearing for slaughter, see *Slaughter v. State*, No. PCD-2004-277 (OK Ct. Crim. App., Jan. 11, 2005), http://caselaw.findlaw.com/ok-court-of-criminal-appeals/1128130.html.

20. Lawrence A. Farwell, "Brain Fingerprinting: A Comprehensive Tutorial Review of Detection of Concealed Information with Event-related Brain Potentials," *Cognitive Neurodynamics* 6 (2012): 115–154, 115; Farwell, "Farwell Brain Fingerprinting: A New Paradigm in Criminal Investigations," self-published paper, Human Brain Research Laboratories, Inc., January 12, 1999, http://www.raven1.net/mcf/bf.htm; Farwell, "Brain Fingerprinting: Brief Summary of the Technology," Brain Fingerprinting Laboratories, 2000, http://www.forensicevidence.com/site/Behv_Evid/Farwell_sum6_00.html. On how memory works, see Daniel Schacter, *The Seven Sins of Memory* (Boston: Houghton Mifflin, 2001); and Edward C. Gooding, "Tunnel Vision: Its Causes and Treatment Strategies," *Journal of Behavioral Optometry* 14, no. 4 (2003): 95–99.

21. Ralf Mertens and John J. B. Allen, "The Role of Psychophysiology in Forensic Assessments: Deception Detection, ERPs, and Virtual Reality Mock Crime Scenarios," *Psychophysiology* 45, no. 2 (2008): 286–298. Allen conducted his own experiment using a mock crime scenario and found that Farwell's method accurately identified "guilty persons" only 50 percent of the time. See John J. B. Allen, "Brain Fingerprinting: Is It Ready for Prime Time?" A Homeland Security Grant from the Office of the Vice President for Research at the University of Arizona per his CV at http://apsychoserver.psychofizz.psych.arizona.edu/JJBAReprints/John_JB_Allen_CV.pdf.

22. Kathleen O'Craven and Nancy Kanwisher, "Mental Imagery of Faces and Places Activates Corresponding Stimulus-Specific Brain Regions," *Journal of Cognitive Neuroscience* 12 (2000): 1013–1023; Jesse Rissman, Henry T. Greely, and Anthony Wagner, "Detecting Individual Memories Through the Neural Decoding of Memory States and Past Experience," *Proceedings of the National Academy of Sciences of the United States of America* 107, no. 21 (2010): 9849–9854. Notably, in both cases—actually seen and believed they had seen—the accuracy of recollection was an anemic accuracy of 59 percent, barely better than chance.

23. British psychiatrist Sean Spence was among the first to suggest using fMRI for this purpose. See Spence et al., "Cognitive Neurobiological Account of Deception," 1755–1762. Depending on the research methodology used, the neural signature of a lie is derived by subtracting information

about subjects' brain activation during the truth state from that during the lie state. In another strategy, researchers compare both the truth and lie conditions with a neutral baseline. Spence et al., "Behavioural and Functional Anatomical Correlates of Deception in Humans," *NeuroReport* 12 (2001): 2849–2853, experimented with thirty people who were questioned about their activities that day (for example, did they make their beds). The subjects took longer (up to 12 percent) to lie than to tell the truth.

24. F. Andrew Kozel et al., "Detecting Deception Using Functional Magnetic Imaging," *Biological Psychiatry* 58, no. 8 (2005): 605–613. For a bibliography of all fMRI lie-detection studies, roughly two dozen, see Henry T. Greely, "Neuroscience, Mind-Reading and the Law," in *A Primer on Criminal Law and Neuroscience: A Contribution of the Law and Neuroscience Project*, ed. Stephen J. Morse and Adina L. Roskies (New York: Oxford University Press, forthcoming).

25. More specifically, the investigators subtracted the number of voxels that were activated more strongly when the subject told the truth about the theft than when he was truthful in answering the neutral questions. This tally represented what could be called "truth" voxels. To determine the number of "lie" voxels, they performed a parallel subtraction: the number of strongly activated voxels associated with a false response to the neutral questions from those associated with a false response to questions about the theft. If the number of "lie" voxels exceeded the number of "truth" voxels, researchers concluded that the subject was being dishonest. Three of those regions—the anterior cingulate cortex, the orbital frontal cortex, and the inferior frontal cortex—showed the greatest activation relative to the other four during the lie condition.

26. Anthony Wagner, "Can Neuroscience Identify Lies?," in *A Judge's Guide to Neuroscience: A Concise Introduction* (Santa Barbara: University of California, Santa Barbara, 2010), 22, http://www.sagecenter.ucsb.edu/sites/staging.sagecenter.ucsb.edu/files/file-and-multimedia/A_Judges_Guide_to_Neuroscience%5Bsample%5D.pdf; G. T. Monteleone et al., "Detection of Deception Using Functional Magnetic Resonance Imaging: Well Above Chance, Though Well Below Perfection," *Social Neuroscience* 4, no. 6 (2009): 528–538.

27. Alexis Madrigal, "MRI Lie Detection to Get First Day in Court," *Wired*, March 16, 2009, http://www.wired.com/wiredscience/2009/03/noliemri/; Hank T. Greely, personal communication, November 29, 2011.

28. For a comprehensive summary of the Semrau case *United States v. Semrau*, see Frances X. Shen and Owen D. Jones, "Brain Scans as Evidence: Truths, Proofs, Lies, and Lessons," *Mercer Law Review* 62 (2011): 861–883. In the 1993 case *Daubert v. Merrell Dow Pharmaceuticals*, the Supreme Court agreed on the several guidelines for admitting scientific expert testimony. Among them are the following factors considered relevant for establishing the "validity" of scientific testimony: (1) empirical testing: whether the theory or technique is falsifiable, refutable, and/or testable; (2) whether it has been subjected to peer review and publication; (3) the known or potential error rate; (4) the existence and maintenance of standards and controls concerning its operation; and (5) the degree to which the theory and the technique are generally accepted by a relevant scientific community. *Daubert v. Merrell Dow Pharmaceuticals*, 509 U.S. 579 (1993).

29. See Greg Miller, "Can Brain Scans Detect Lying?," May 14, 2010, http://news.sciencemag.org/scienceinsider/2010/05/can-brain-scans-detect-lying-exc.html; and Alexis Madrigal, "Eyewitness Account of 'Watershed' Brain Scan Legal Hearing," *Wired*, May 17, 2010, http://www.wired.com/wiredscience/2010/05/fmri-daubert/#more-21661.

30. Alexis Madrigal, "Brain Scan Evidence Rejected by Brooklyn Court," *Wired*, May 5, 2010, http://www.wired.com/wiredscience/2010/05/fmri-in-court-update/; Michael Laris, "Debate on Brain Scans as Lie Detectors Highlighted in Maryland Murder Trial," *Washington Post*, August 26, 2012.

31. Nancy Kanwisher, "The Use of fMRI in Lie Detection: What Has Been Shown and What Has Not," in *Using Imaging to Identify Deceit: Scientific and Ethical Questions* (Cambridge, MA: American Academy of Arts and Sciences, 2009), 7–13, 12. Hank Greely, professor of law at Stan-

ford University, playfully offers one way to meet the challenge: "Take undergraduates, randomly arrest them, accuse them of something you think they may be guilty of, scare them into thinking they're actually going to be imprisoned and wait to see whether you can detect the lies they can tell." But, he adds, no research review board would allow that experiment. An even more unlikely scenario entails having actual suspects serve as subjects. A third party privy to solid forensic evidence would know before testing who was innocent and who was not. The suspects would need to believe that the disposition of their cases hangs on the fMRI results. Hank Greely, personal communication, May 12, 2010. On movement reducing the accuracy of lie detection, see Giorgio Ganis et al., "Lying in the Scanner: Covert Countermeasures Disrupt Deception Detection by Functional Magnetic Resonance Imaging," *Neuroimage* 55, no. 1 (2011): 312–319. The movement affected signals from the lateral and medial prefrontal cortices—signals that would otherwise have differentiated deceptive from honest responses—to the point that the differential activation became much less pronounced, and overall accuracy was considerably degraded.

32. Elizabeth A. Phelps, "Lying Outside the Laboratory: The Impact of Imagery and Emotion on the Neural Circuitry of Lie Detection," in *Using Imaging to Identify Deceit*, 14–22. See also Daniel L. Schacter and Scott D. Slotnick, "The Cognitive Neuroscience of Memory Distortion," *Neuron* 44 (2004): 149–160.

33. Joseph Henrich, Steven J. Heine, and Ara Norenzayan, "The Weirdest People in the World?," *Behavioral and Brain Sciences* 33, nos. 2–3 (2010): 61–83, contrast the Americans who typically wind up as psychology subjects with the whole population of the United States and highlight the diversity among adult Americans in such areas as social behavior, moral reasoning, cooperation, fairness, performance on IQ tests, and analytical abilities. U.S. undergraduates exhibit demonstrable differences not only from Americans who are not university educated but also from previous generations of their own families. On experienced liars, see B. Verschuere et al., "The Ease of Lying," *Consciousness and Cognition* 20, no. 3 (2011): 908–911. On neural activity during thinking about lying and during lying, see Joshua D. Greene and Joseph M. Paxton, "Patterns of Neural Activity Associated with Honest and Dishonest Moral Decisions," *Proceedings of the National Academy of Sciences* 106, no. 30 (2009): 12506–12511.

34. Spence et al., "Cognitive Neurobiological Account of Deception"; *Philosophical Transactions of The Royal Society of London B: Biological Sciences* 359 (2004). Spence et al., "Behavioral and Functional Anatomical Correlates of Deception in Humans," *Neuroreport* 12 (2001); 2349–2353. See also G. Ganis et al., "Visual Imagery in Cerebral Visual Dysfunction," *Neorologic Clinics of North America* 21 (2003): 631–646; and Henry T. Greely and Judy Illes, "Neuroscience-Based Lie Detection: The Urgent Need for Regulation," *American Journal of Law and Medicine* 33 (2007): 377–431. Doubtless, lying is a behavior that recruits functions of many brain regions, but it is easy to fall into the reverse-inference trap and assume that areas differentially activated during lying are necessarily the only ones involved in deception. See Anthony Wagner, "Can Neuroscience Identify Lies?," 20.

35. Giorgio Ganis et al., "Neural Correlates of Different Types of Deception: An fMRI Investigation," *Cerebral Cortex* 13 (2003): 830–836.

36. Ahmed A. Karim et al., "The Truth About Lying: Inhibition of the Anterior Prefrontal Cortex Improves Deceptive Behavior," *Cerebral Cortex* 20, no. 1 (2010): 205–213; Stephen M. Kosslyn, "Brain Bases of Deception: Why We Probably Will Never Have a Perfect Lie Detector," Berkman Center for Internet and Society at Harvard University, January 11, 2010, http://cyber.law .harvard.edu/events/lawlab/2010/01/kossyln.

37. Margaret Talbot, "Duped: Can Brain Scans Uncover Lies?," *New Yorker,* July 2, 2007, http://www.newyorker.com/reporting/2007/07/02/070702fa_fact_talbot; Frederick Schauer, "Can Bad Science Be Good Evidence? Neuroscience, Lie Detection, and Beyond," *Cornell Law Review* 95 (2010): 1190–1220, 1194.

38. Giorgio Ganis and Julian Paul Keenan, "The Cognitive Neuroscience of Deception," *Social Neuroscience* 4, no. 6 (2009): 465–472. Emotions, such as guilt and anxiety, will probably

modulate how quickly and efficiently the deceiver can process these tasks. Nonetheless, fMRI is a useful tool for examining the neural correlates of lying, and some investigators have begun pairing it with other modalities, such as EEG and transcranial magnetic stimulation (TMS). Bruce Luber et al., "Non-invasive Brain Stimulation in the Detection of Deception: Scientific Challenges and Ethical Consequences," *Behavioral Sciences and the Law* 27, no. 2 (2009): 191–208, http://www .scribd.com/doc/13112142/Noninvasive-brain-stimulation-in-the-detection-of-deception-Scientific -challenges-and-ethical-consequences.

39. In 2009, a National Research Council review committee uniformly agreed that "to date, insufficient, high-quality research has been conducted to provide empirical support for the use of any single neurophysiological technology to detect deception." National Research Council's Committee on Military and Intelligence Methodology for Emergent Neurophysiological and Cognitive/ Neural Science Research in the Next Two Decades, "Emerging Cognitive Neuroscience and Related Technologies," 4, http://books.nap.edu/openbook.php?record_id=12177&page=4. In a 2008 report, the NRC concluded that "opinions differed within the committee concerning the 'near-term contribution of functional neuroimaging to the development of a system to detect deception in a practical or forensic sense.'" National Research Council, "Opportunities in Neuroscience for Future Army Applications," NRC 200996, 4, http://www.nap.edu/catalog.php?record_id=12500. Cephos was established in 2004 but took on clients in 2008; see www.cephoscorp.com/about-us /index.php. Dr. Kozel is a nonpaid adviser to Cephos, and his university, Medical University of South Carolina, has a patent agreement with Cephos; http://www.cephoscorp.com/about-us/index .php#scientific. The University of Pennsylvania licensed the pending patents on his research to No Lie MRI in 2003, based on Daniel Langleben's work, in exchange for an equity position in the company; see Lee Nelson, "The Inside Image," *Advanced Imaging*, September 2008, 8–11. Huizenga is quoted in Mark Harris, "MRI Lie Detectors," *IEEE Spectrum*, August 2010, http://spectrum .ieee.org/biomedical/imaging/mri-lie-detectors/0. Veritas Scientific is a division of No Lie MRI that focuses on developing implementation and application of No Lie MRI software for use by the U.S. military, government agencies, law-enforcement agencies, and foreign governments; see http:// noliemri.com/investors/Overview.htm; emphasis added.

40. Via e-mail from Mr. Nathan dated November 14, 2011. "Neuroscientist Uses Brain Scan to See Lies Form," *National Public Radio*, October 30, 2007, http://www.npr.org/templates/transcript /transcript.php?storyId=15744871. Huizenga is quoted at http://www.cephoscorp.com/about-us /index.php. In a personal communication with Harvey Nathan, February 26, 2010, he told us that he had contacted numerous university researchers to ask if they would conduct additional fMRI tests for him, hoping that repeated tests would sway insurance companies to pay up. "No one was willing to do it," he reported; "they said it wasn't ready to test commercially."

41. David Washburn, "Can This Machine Prove If You're Lying?," *Voice of San Diego*, April 2, 2009, http://m.voiceofsandiego.org/mobile/science/article_bcff9425-cae5-5da4-b036-3dbdc0e82d5e .html; Greg Miller, "fMRI Lie Detection Fails a Legal Test," *Science* 328 (2010): 1336–1337.

42. John Ruscio, "Exploring Controversies in the Art and Science of Polygraph Testing," *Skeptical Inquirer* 29 (2005): 34–39; transcript of a conversation between President Richard M. Nixon, John D. Ehrlichman, and Egil Krogh Jr. on July 24, 1971, 5 (on file with the authors). In the Watergate tapes, Nixon expressed concern over leaks regarding the Strategic Arms Limitation Treaty (SALT) talks. He spoke of subjecting hundreds of government employees to pinpoint the source of leaks regarding an international treaty and opined that perhaps unsavory characters would be dissuaded from applying for such jobs in the first place if they knew that they would be subjected to taking a lie-detector test. On research on the effect of fake lie-detector equipment, see Saul Kassin, Steven Fein, and Hazel Rose Markus, *Social Psychology*, 7th ed. (Boston: Houghton Mifflin, 2007); and Theresa A. Gannon, Kenneth Keown, and D. L. Polaschek, "Increasing Honest Responding on Cognitive Distortions in Child Molesters: The Bogus Pipeline Revisited," *Sexual Abuse: A Journal of Research and Treatment* 19, no. 1 (2007): 5–22.

43. David McCabe, Alan D. Castel, and M. G. Rhodes, "The Influence of fMRI Lie Detection Evidence on Juror Decision Making," *Behavioral Sciences and the Law* 29 (2011): 566–577.

44. Many legal scholars and judges believe that no lie-detection device, regardless of its accuracy, should be allowed in a courtroom. The concept of the jury's role as the ultimate arbiter of witness credibility has long-standing roots in jurisprudence. See *United States v. Scheffer*, 523 U.S. 303 (1998), 312–313 (plurality opinion) (noting that lie-detector evidence "diminish[es]" juries' role as the mechanism by which credibility is assessed); see citations 43–44; *United States v. Call*, 129 F.3d 1402, 1406 (10th Cir. 1997) (ruling that trial court did not abuse its discretion in excluding polygraph evidence under Rule 403 because, inter alia, such testimony "usurps a critical function of the jury and because it is not helpful to the jury, which is capable of making its own determination regarding credibility"); and Julie Seaman, "Black Boxes: FMRI Lie Detection and the Role of the Jury," *University of Akron Law Review* 42 (2009): 931–941. In a 2010 New York case on employment discrimination, a young woman who worked for a temp agency alleged that her formal complaint about sexual harassment by her boss resulted in her getting no assignments. A coworker was said to have overheard their boss discussing just such retaliation against the woman. The coworker took a Cephos-administered lie-detection test and passed, but the court would not allow Steven Laken of Cephos to testify on the results. "Anything that impinges on the province of the jury on issues of credibility should be treated with a great deal of skepticism," the judge wrote in *Wilson v. Corestaff Services, L.P.*, 28 Misc. 3d 428 (Supreme Court, Kings County, 2010), http://www.courts.state.ny.us/reporter/3dseries/2010/2010_20176.htm. See also Grace West, "Brooklyn Lawyer Seeks to Use Brain Scan as Lie Detector in Court," *NBC New York*, May 5, 2010, http://www.nbcnewyork.com/news/local-beat/Brain-scanning-92888084.html. The American Civil Liberties Union has registered concern about what could be the ultimate violation of privacy and other watch-groups advocating what they call "cognitive liberty." See ACLU Press Release, "ACLU Seeks Information About Government Use of Brain Scanners in Interrogations," June 28, 2006, http://www.aclu.org/technology-and-liberty/aclu-seeks-information-about-government-use-of-brain-scanners-interrogations. "And we are still in our infancy when it comes to understanding the underlying processes of the brain that the scanners have begun to reveal. We do not want to see our government yet again deploying a potentially momentous technology unilaterally and in secret, before Americans have had a chance to figure out how it fits in with our values as a nation." For the quotation from the ACLU spokesperson, see Jay Stanley, "High-Tech 'Mind Readers' Are Latest Effort to Detect Lies," August 29, 2012, http://www.aclu.org/blog/technology-and-liberty/high-tech-mind-readers-are-latest-effort-detect-lies. "Mental privacy panic": Francis X. Shen, "Neuroscience, Mental Privacy, and the Law," 36 *Harvard Journal of Law and Public Policy* (forthcoming April 2013). On regulation of the technology, see Henry T. Greely, "Premarket Approval for Lie Detections: An Idea Whose Time May Be Coming," *American Journal of Bioethics* 5(2005): 50–52. Jonathan Moreno, *Mind Wars: Brain Science and the Military in the 21st Century* (New York: Bellevue Literary Press, 2012), 186, calls for the creation of a national advisory committee on neurosecurity, staffed by professionals who possess the relevant scientific, ethical, and legal expertise. The committee would be analogous to the National Science Advisory Board for Biosecurity, which was established in 2004 and is administered by the National Institutes of Health but advises all cabinet departments on how to minimize the misuse of biological research.

45. S. E. Stoller and P. R. Wolpe, "Emerging Neurotechnologies for Lie Detection and the Fifth Amendment," *American Journal of Law and Medicine* 33 (2007): 3359–3375; Amanda C. Pustilnik, "Neurotechnologies at the Intersection of Criminal Procedure and Constitutional Law," in *The Constitution and the Future of Criminal Law*, ed. Song Richardson and John Parry (Cambridge: Cambridge University Press, forthcoming); Michael S. Pardo, "Disentangling the Fourth Amendment and the Self-Incrimination Clause," *Iowa Law Review* 90, no. 5 (2005): 1857–1903.

46. Nita Farahany, "Incriminating Thoughts," *Stanford Law Review* 64 (2012): 351–408. Farahany proposes an alternative approach to classifying evidence subject to the privilege against

self-incrimination, which focuses on the information's function rather than its form. She introduces an alternative categorization—identifying, automatic, memorialized, or uttered evidence—a spectrum of evidence on which emerging neuroscience could more easily align.

47. Nita A. Farahany, "Searching Secrets," *University of Pennsylvania Law Review* 160 (2012): 1239–1308 (discussing Fourth Amendment implications of emerging neurotechnologies); Robin G. Boire, "Searching the Brain: The Fourth Amendment Implications of Brain-Based Deception Detection Devices," *American Journal of Bioethics* 5, no. 2 (2005): 62–63.

Chapter 5

1 See *Roper v. Simmons*, 543 U.S. 551 (2005), http://www.law.cornell.edu/supct/html/03-633 .ZO.html.

2. Oral argument of Mr. James W. Ellis in *Atkins v. Virginia*, 536 U.S. 304 (2002); see transcript at http://www.oyez.org/cases/2000–2009/2001/2001_00_8452/.

3. In May 2002, the U.S. Supreme Court decided six to three in favor of Atkins. The justices ruled that because of "disabilities in areas of reasoning, judgment, and control of their impulses, [the mentally retarded] do not act with the level of moral culpability that characterizes the most serious adult criminal conduct." See *Atkins*, 536 U.S. 304, http://www.law.cornell.edu/supct/html /00-8452.ZO.html. For the argument of Simmons's counsel, see http://www.internationaljusti ceproject.org/pdfs/SimmonsAtkinsbrief-final.pdf. On August 26, 2003, the Missouri Supreme Court vacated the death sentence, holding that juvenile executions violate the Eighth Amendment of the U.S. Constitution under the "evolving standards of decency." Brain science did not seem to impress them: "While the parties have cited this Court to numerous current studies and scientific articles about the structure of the human mind . . . this Court need not look so far afield." See http:// caselaw.findlaw.com/mo-supreme-court/1273234.html.

4. Transcript of oral argument at http://www.oyez.org/cases/2000–2009/2004/2004_03_633. Over four hundred health care experts showed support by signing a "Health Professionals' Call to Abolish the Execution of Juvenile Offenders in the United States," which stated, "It is unfair and unreasonable to impose expectations of adult-level capacities on the thinking and behavior of minors"; http://www.hrea.org/lists/psychology-humanrights-l/markup/msg00364.html. Carolyn Y. Johnson, "Brain Science v. the Death Penalty," *Boston Globe*, October 12, 2004, http://www .boston.com/news/globe/health_science/articles/2004/10/12/brain_science_v_death_penalty/.

5. See "Brief of the American Medical Association, American Psychiatric Association, American Society for Adolescent Psychiatry, American Academy of Child and Adolescent Psychiatry, American Academy of Psychiatry and the Law, National Association of Social Workers, Missouri Chapter of the National Association of Social Workers, and National Mental Health Association as Amici Curiae in Support of Respondent," 2005, http://www.ama-assn.org/resources/doc/legal-issues /roper-v-simmons.pdf.

6. Ibid.; the amici cited, among others, Elizabeth R. Sowell et al., "Mapping Continued Brain Growth and Gray Matter Density Reduction in Dorsal Frontal Cortex: Inverse Relationships During Post-adolescent Brain Maturation," *Journal of Neuroscience* 21, no. 22 (2001): 8819–8829; and Laurence Steinberg and Elizabeth S. Scott, "Less Guilty by Reason of Adolescence: Developmental Immaturity, Diminished Responsibility, and the Juvenile Death Penalty," *American Psychologist* 58, no. 12 (2003): 1009–1018.

7. According to the amici, myelination and pruning were once thought to be complete by puberty, but techniques developed over the past two decades show that both persist into the mid-twenties. Since the briefs were written, the picture has become even more complicated. For example, some studies have shown that dangerous teens are actually more likely than their counterparts to have especially well-myelinated, more mature, tracts coursing from the prefrontal cortex to other

parts of the brain; see Gregory S. Berns, Sara Moore, and C. Monica Capra, "Adolescent Engagement in Dangerous Behaviors Is Associated with Increased White Matter Maturity of Frontal Cortex," *PLoS One* 4, no. 8 (2009): e6773, doi:10.1371/journal.pone.0006773. On the adolescent reward system, see Adriana Galvan et al., "Earlier Development of the Accumbens Relative to Orbitofrontal Cortex Might Underlie Risk-Taking Behavior in Adolescents," *Journal of Neuroscience* 26, no. 25 (2006): 6885–6892; and Matthew J. Fuxjager et al., "Winning Territorial Disputes Selectively Enhances Androgen Sensitivity in Neural Pathways Related to Motivation and Social Aggression," *Proceedings of the National Academy of Sciences* 107 (2010): 12393, 12396. For the quotation from the amicus brief, see http://www.abanet.org/crimjust/juvjus/simmons/ama.pdf (see p. 22 of website for more information).

8. Jeffrey Rosen, "The Brain on the Stand," *New York Times Magazine*, March 11, 2007, http://www.nytimes.com/2007/03/11/magazine/11Neurolaw.t.html?pagewanted=1&_r=1&ref=science. For the *Roper v. Simmons* ruling, see www.supremecourt.gov/opinions/04pdf/03-633.pdf.

9. In its initial use, in the 1900s, the term "neurolaw" meant something different. A personal injury lawyer coined the term to signify the growing importance of testimony by brain experts in traumatic-brain-injury cases. S. J. Taylor, "Neurolaw: Towards a New Medical Jurisprudence," *Brain Injury* 9, no. 7 (1995): 745–751; and Owen D. Jones and Francis X. Shen, "Law and Neuroscience in the United States," in *International Neurolaw*, ed. Tade Spranger (Berlin: Springer-Verlag, 2012), 349–380. In 2011, it was renewed for $4.85 million; see Amy Wolf, "Landmark Law and Neuroscience Network Expands at Vanderbilt," *Research News at Vanderbilt*, August 24, 2011, http://news.vanderbilt.edu/2011/08/grant-will-expand-law-neuroscience-network/; it was led by Owen Jones, who stated in 2007, "Neuroscientists need to understand law, and lawyers need to understand neuroscience," in http://www.macfound.org/press/press-releases/new-10-million-macarthur-project-integrates-law-and-neuroscience/. Stephen Morse spoke before President Bush's Council on Bioethics; see http://bioethics.georgetown.edu/pcbe/transcripts/sep04/session1.html. Morse and Martha Farah spoke before President Obama's Bioethics Commission in Washington, D.C., February 28–March 1, 2011; for Martha Farah's remarks, see http://www.tvworldwide.com/events/bioethics/110228/globe_show/default_go_archive.cfm?gsid=1552&type=flv&test=0&live=0; for Stephen Morse's remarks, see http://www.tvworldwide.com/events/bioethics/110228/globe_show/default_go_archive.cfm?gsid=1546&type=flv&test=0&live=0. For the Royal Society's attention to these issues, see *Brain Waves Module 4: Neuroscience and the Law* (London: Royal Society, 2011), http://royalsociety.org/uploadedFiles/Royal_Society_Content/policy/projects/brain-waves/Brain-Waves-4.pdf. See http://www.lawneuro.org/bibliography.php for an ongoing, updated bibliography maintained by Francis X. Shen of the University of Minnesota Law School. Examples of blogs are http://lawneuro.typepad.com/the-law-and-neuroscience-blog/neurolaw/ (The MacArthur Foundation Research Network on Law and Neuroscience); http://blogs.law.stanford.edu/lawandbiosciences/ (The Center for Law and Biosciences, Stanford Law School); and http://kolber.typepad.com/ (Adam Kolber, professor, Brooklyn Law School). Some law schools with courses or lecture series are the University of Maryland, Tulane, Stanford, the University of Pennsylvania, the University of Akron, the University of Arizona, Harvard, the University of California at Berkeley, and Brooklyn Law School. The Baylor College of Medicine has an Initiative on Neuroscience and Law.

10. "Some sort of organic brain defense has become de rigueur in any sort of capital defense," according to Daniel Martell, head of Forensic Neuroscience Consultants, Inc., in Rosen, "Brain on the Stand." "A number of judges have informally told us that evidence from neuroscience, including fMRI, has become standard in capital sentencing"; Walter Sinnott-Armstrong et al., "Brain Images as Legal Evidence," *Episteme* 5, no. 3 (2008): 359–373, 369, http://muse.jhu.edu/journals/epi/summary/v005/5.3.sinnott-armstrong.html. For appeals of convictions, see *Lathram v. Johnson*, 2011 WL 676962 (E.D. Va. 2011); *People v. Jones*, 620 N.Y.S.2d 656 (App. Div. 1994), aff'd, 85 N.Y.2d 998 (1995); and Shamael Haque and Melvin Guyer, "Neuroimaging Studies in Diminished Capacity Defense," *Journal of the American Academy of Psychiatry and the Law* 38, no. 4

(2010): 605–607, http://www.jaapl.org/cgi/content/full/38/4/605. Faigman is quoted in Lizzie Buchen, "Science in Court: Arrested Development," *Nature* 484 (2012): 304–306, at 306.

11. Federal Insanity Defense Reform Act, 18 U.S.C. § 17 (1984).

12. Stephen J. Morse, "Brain Overclaim Syndrome and Criminal Responsibility: A Diagnostic Note," *Ohio State Journal of Criminal Law* 3 (2006): 397–412, at 399. Morse, "Inevitable Mens Rea," *Harvard Journal of Law and Public Policy* 27, no. 1 (2003): 51–64.

13. "Dugan's case became what is thought to be the first in the world to admit fMRI as evidence"; Virginia Hughes, "Science in Court: Head Case," *Nature* 464 (2010): 340–342, http://www.nature .com/news/2010/100317/full/464340a.html. On the moral disability of psychopaths, see Richard E. Redding, "The Brain-Disordered Defendant: Neuroscience and Legal Insanity in the Twenty-First Century," *American University Law Review* 56 (2006): 51–126. See also Maaike Cima, Franca Tonnaer, and Marc D. Hauser, "Psychopaths Know Right from Wrong but Don't Care," *Social Cognitive and Affective Neuroscience* 5 (2010): 59–67; and Andrea L. Glenn, "Moral Decision Making and Psychopathy," *Judgment and Decision Making*, vol. 5, no. 7 (2010): 497–505.

14. Robert D. Hare, "Psychopathy, Affect and Behaviour," in *Psychopathy: Theory, Research and Implications for Society*, ed. D. J. Cooke, Adelle E. Forth, and Robert D. Hare (Dordrecht: Kluwer, 1998), 105–139.

15. Robert D. Hare et al., "Psychopathy and the Predictive Validity of the PCL-R: An International Perspective," *Behavioral Sciences and the Law* 18, no. 5 (2000): 623–645 (38.5 was an average score). The brain regions that are involved include the amygdala, the orbitofrontal cortex, the insula, the cingulate cortex, and the ventromedial prefrontal cortex. See Adrian Raine and Yaling Yang, "Neural Foundations to Moral Reasoning and Antisocial Behavior," *Social Cognitive and Affective Neuroscience* 1, no. 3 (2006): 203–213; Andrea L. Glenn and Adrian Raine, "The Neurobiology of Psychopathy," *Psychiatric Clinics of North America* 31, no. 3 (2008): 463–475; and R. J. R. Blair, "The Amygdala and Ventromedial Prefrontal Cortex: Functional Contributions and Dysfunction in Psychopathy," *Philosophical Transactions of the Royal Society of London B: Biological Sciences* 363, no. 1503 (2008): 2557–2565.

16. Kent A. Kiehl and Joshua W. Buckholtz, "Inside the Mind of a Psychopath," *Scientific American Mind*, September/October 2010, 22–29; Carla L. Harenski et al., "Aberrant Neural Processing of Moral Violations in Criminal Psychopaths," *Journal of Abnormal Psychology* 119, no. 4 (2010): 863–874.

17. Hughes, "Science in Court," 342.

18. Nicole Rafter, "The Murderous Dutch Fiddler: Criminology, History and the Problem of Phrenology," *Theoretical Criminology* 9, no. 1 (2005): 65–96, 86, http://www.sagepub.com/tibbetts /study/articles/SectionIII/Rafter.pdf. Stacey A. Tovino, "Imaging Body Structure and Mapping Brain Function: A Historical Approach," *American Journal of Law and Medicine* 33 (2007), 193–228; John Van Wyhe, "The Authority of Human Nature: The Schädellehre of Franz Joseph Gall," *British Journal for the History of Science* 35, no. 124, pt. 1 (2002): 17–42, discusses Gall's work in prisons and with criminals to develop and perfect his cranial measurements and trait localizations.

19. Cesare Lombroso, *Criminal Man*, summarized by Gina Lombroso-Ferrero (New York: Knickerbocker Press, 1911), 6, http://www.gutenberg.org/files/29895/29895-h/29895-h.htm. The Italian physician, who remained an adherent of phrenology even after it fell into disrepute by the middle of the nineteenth century, studied the physical attributes of prison inmates. Lombroso described how their skulls bore primitive animal-like features, such as a sloping forehead, high cheekbones, and large eye sockets. Their brutish instincts were untamed by the inborn power of self-control possessed by civilized men and women. Lombroso's work echoed Greek physiognomy, the putative notion of a correspondence between the features of the face and a person's inner self, but also looked to the brain, albeit indirectly. Nicole H. Rafter, *The Criminal Brain: Understanding Biological Theories of Crime* (New York: New York University Press, 2008). The quotation is from Lombroso, *Criminal Man*, 6, as cited in Stephen Jay Gould, *Ontogeny and Phylogeny* (Cambridge,

MA: Belknap Press of Harvard University Press, 1977), 122. For a discussion of Lombroso's theories, see Helen Zimmern, "Reformatory Prisons and Lombroso's Theories," *Popular Science Monthly* 43 (1893): 598–609.

20. Vernon Mark, William Sweet, and Frank Ervin, "The Role of Brain Disease in Riots and Urban Violence," *Journal of the American Medical Association* 201, no. 11 (1967): 895; Vernon H. Mark and Frank R. Ervin, *Violence and the Brain* (New York: Harper and Row, 1970). On page 33, the authors speculate, "Either the limbic system has become pathologically hyperactive or its neocortical ([frontal] control) inputs have become abnormal." On public concern, see Leroy Aarons, "Brain Surgery Is Tested on 3 California Convicts," *Washington Post*, February 25, 1972; and Lori Andrews, "Psychosurgery: The New Russian Roulette," *New York Magazine*, March 7, 1977, 38–40. The term "identity-destroying" is used in a March 21, 1977, letter to the editor, *New York Magazine*, 7. Congressional hearing: Bertram S. Brown from *Report and Recommendations: Psychosurgery: The National Commission for the Protection of Human Subjects of Biomedical and Behavioral Research* (Washington, DC: Department of Health and Welfare, 1977), 10, http://video cast.nih.gov/pdf/ohrp_psychosurgery.pdf. The commission concluded that a ban on the performance of all psychosurgical procedures was not an appropriate response to instances of misuse.

21. Jonathan Brodie, MD, witness for the prosecution, argued that it was not possible to draw conclusions about psychopathic behavior based on fMRI data, much less about behavior that took place twenty-six years earlier; see Barbara Bradley Hagerty, "Inside a Psychopath's Brain: The Sentencing Debate," *National Public Radio*, June 30, 2010, http://www.npr.org/templates/story /story.php?storyId=128116806. Kiehl did not scan individuals who scored very high on the Hare Psychopathy Checklist but did not show similar activity pattern; this would be an important comparison group. See "Can Genes and Brain Abnormalities Create Killers?," *National Public Radio*, July 6, 2010, http://www.npr.org/templates/story/story.php?storyId=128339306; and Mehmet K. Mahmut, Judi Homewood, and Richard Stevenson, "The Characteristics of Non-criminals with High Psychopathy Traits: Are They Similar to Criminal Psychopaths?," *Journal of Research in Personality* 42, no. 3 (2008): 679–692. These investigators found that noncriminal psychopaths, despite exhibiting the same neuropsychological profile as criminal psychopaths, are somehow protected from developing antisocial tendencies, such as emotional and financial havoc, perhaps by good parental attachments, which may steer biological underpinnings.

22. Joseph H. Baskin, Judith G. Edersheim, and Bruce H. Price, "Is a Picture Worth a Thousand Words? Neuroimaging in the Courtroom," *American Journal of Law and Medicine* 33 (2007): 239–269; Teneille Brown and Emily Murphy, "Through a Scanner Darkly: Functional Neuroimaging as Evidence of a Criminal Defendant's Past Mental States," *Stanford Law Review* 62, no. 4 (2010): 1119–1208.

23. Zoe Morris et al., "Incidental Findings on Brain Magnetic Resonance Imaging: Systematic Review and Meta-Analysis," *British Medical Journal* 339 (2009), http://www.bmj.com/highwire /filestream/386096/field_highwire_article_pdf/0/bmj.b3016. Damage to underlying axons may go undetected structurally, but these axons may still be capable of altering behavior. See Susumu Mori and Jiangyang Zhang, "Principles of Diffusion Tensor Imaging and Its Applications to Basic Neuroscience Research," *Neuron* 51, no. 5 (2005): 527–539.

24. *People v. Weinstein*, 591 N.Y.S.2d 715 (Sup. Ct. 1992), at http://www.leagle.com/xmlResult .aspx?xmldoc=1992190156Misc2d34_1186.xml&docbase=CSLWAR2-1986-2006.

25. *People v. Weinstein*, 591 N.Y.S.2d 715 (Sup. Ct. 1992), at 717–718, 722–723. The court admitted the PET scan results "despite evidence that such pathology has no known link to criminal behavior" but prohibited testimony about whether an arachnoid cyst or metabolic disturbances "in the frontal lobes of the brain directly cause violence."

26. J. Rojas-Burke, "PET Scans Advance as Tool in Insanity Defense," *Journal of Nuclear Medicine* 34, no. 1 (1993): 13N–26N, 13N, 16N (quoting Dr. Jonathan Brodie). Furthermore, a cyst disruptive enough to obliterate his moral sense, the experts concluded, would most likely have

affected him in other ways: headaches on the front-left side, for example, or problems with low frustration tolerance, impulsiveness, aggression, or problem solving.

27. Brian W. Haas and Turhan Canli, "Emotional Memory Function, Personality Structure and Psychopathology: A Neural System Approach to the Identification of Vulnerability Markers," *Brain Research Reviews* 58, no. 1 (2008): 71–84.

28. Jeffrey M. Burnsand and Russell H. Swerdlow, "Right Orbitofrontal Tumor with Pedophilia Symptom and Constructional Apraxia Sign," *Archives of Neurology* 60, no. 3 (2003): 437–440. The patient was convicted of child molestation and sentenced to attend a program for sex offenders, where he propositioned staff members, fully knowing that he risked prison time for doing so. Shortly thereafter, he developed headaches and other neurological symptoms, and finally, the tumor was detected.

29. Charles Montaldo, "The Call to Police: The Andrea Yates Case," *About.com*, http://crime .about.com/od/female_offenders/a/call_yates.htm. Yates is quoted in Timothy Roche, "Andrea Yates: More to the Story," *Time*, March 18, 2002. http://www.time.com/time/nation/article/0,8599 ,218445-1,00.html; and "A Dark State of Mind," *Newsweek*, March 3, 2002, http://www.thedai-lybeast.com/newsweek/2002/03/03/a-dark-state-of-mind.html.

30. In 2002, Yates was convicted of murder for three of the deaths of her five children. Narrowly speaking, Yates intentionally killed her children, but her reason was severely corrupted, in our view, to the point of legal insanity. Strictly speaking, she was able to form the intent to perform an act she knew was legally wrong—that is, she possessed *mens rea*—but her severe postpartum psychosis prevented her from correctly grasping the moral significance of her deeds. On her retrial, see "Yates Retrial May Signal Opinion Shift," *USA Today*, July 72, 2006, http://www.usatoday .com/news/nation/2006–07-27-yates-verdict_x.htm. It is also possible that public opinion about the role of mental illness evolved over the period between the crime and her retrial. See also "Insanity Plea Successful in Andrea Yates Retrial," *Psychiatric News* 41, no. 16 (2006): 2–3.

31. Martha J. Farah and Seth J. Gillihan, "The Puzzle of Neuroimaging and Psychiatric Diagnosis: Technology and Nosology in an Evolving Discipline," *American Journal of Bioethics Neuroscience* 3, no. 4 (2012): 1–11.

32. Available data from Nita Farahany, cited in *Brain Waves Module 4*. Figure 1 (p. 4) showed that from 2005 to 2009, the number of criminal cases in which neurological or behavioral genetic evidence was introduced doubled from 101 to 205. Figure 3 (p. 19) is based on cases in the United States in which neuroscientific or genetic evidence is discussed in a judicial opinion. From 843 opinions (majority, plurality, concurrence, dissent), and 722 unique cases analyzed between 2004 and 2009, the introduction of such evidence most commonly occurred in first-degree murder cases (*n* = 449 cases), homicide (91), rape (54) and "other" (222), and robbery (174), and least in felony murder (36), child abuse (33), and carjacking (15). Stephen Morse, in a personal communication to the authors, August 27, 2012, stated, "Anecdotally, there is belief that NS claims are being made more often." "Efforts to bring neuroscientific evidence into the courtroom have definitely increased"; Greg Miller, Science Podcast, September 15, 2011, with Owen Jones and Martha Farah, at minute 3:07, http://news.sciencemag.org/sciencenow/2011/09/live-chat-brain-science-and-the .html. Helen S. Mayberg is a neurologist at Emory University and has testified in many trials. She spoke about the burgeoning courtroom use of imaging studies at a presentation to MacArthur Foundation on Law and Neuroscience members at its annual meeting in Santa Barbara, California, on May 29, 2008, as cited in Brown and Murphy, "Through a Scanner Darkly." Recent courtroom uses of imaging technology are discussed in Purvak Patel et al., "The Role of Imaging in United States Courtrooms," *Neuroimaging Clinics of North America* 17, no. 4 (2007): 557–567. On the federal level, admissibility standards are high, but in most capital cases the threshold is low. The *Daubert* standard is a rule of evidence regarding the admissibility of expert witness testimony during U.S. federal legal proceedings. In *Daubert v. Merrill Dow Pharmaceuticals*, 509 U.S. 579 (1993) the Court held that federal judges have a duty to "ensure that any and all scientific testi-

mony or evidence admitted is not only relevant, but reliable." "Reliable" requires that the following questions be answered: (1) Has the technique been subjected to falsification and refutation via experimentation? (2) Has the technique been subjected to peer review and publication? (3) What is the known or potential rate of error? (4) Has the technique been generally accepted within the relevant scientific community? In addition to reliability, under Federal Rule of Evidence 403, judges are also directed to consider any potential prejudicial impact of scientific evidence in trials. See http://www.law.cornell.edu/rules/fre/rule_403.

33. Rosen, "Brain on the Stand."

34. Ken Strutin, "Neurolaw: A New Interdisciplinary Research," *New York Law Journal*, January 15, 2009, http://www.law.com/jsp/lawtechnologynews/PubArticleLTN.jsp?id=1202427455426 &Neurolaw_A_New_Interdisciplinary_Research&slreturn=20130026161430. Some legal scholars deem brain scans so "prejudicial" that they have called for their exclusion altogether or at least for a moratorium on their use. Jane Campbell Moriarty, "Flickering Admissibility: Neuroimaging Evidence in the U.S. Courts," *Behavioral Sciences and the Law* 26, no. 1 (2008): 29–48, esp. 48; Brown and Murphy, "Through a Scanner Darkly," 1188–1202. For a more liberal view, see Adam Teitcher, "Weaving Functional Brain Imaging into the Tapestry of Evidence: A Case for Functional Neuroimaging in Federal Criminal Courts," *Fordham Law Review* 80, no. 1 (2011): 356–401.

35. Madeleine Keehner, Lisa Mayberry, and Martin H. Fischer, "Different Clues from Different Views: The Role of Image Format in Public Perception of Neuroimaging Results," *Psychonomic Bulletin and Review* 18, no. 2 (2011): 422–428.

36. David P. McCabe and Alan D. Castel, "Seeing Is Believing: The Effect of Brain Images on Judgments of Scientific Reasoning," *Cognition* 107 (2008): 343–352. Some critics have correctly pointed out that McCabe and Castel's research design was not perfect: The brain images they presented to participants featured identical patterns of activation in the same brain regions across the two conditions and therefore contained more information than the bar graphs, which depicted only the total activations in these brain regions. We do not find this criticism fatal to McCabe and Castel's central thesis, however. The added information afforded by the brain images—above and beyond the bar graph—was logically irrelevant to the flawed explanations the subjects had to evaluate; it indicated only correlation, not causation. Fancy brain images notwithstanding, a causal inference linking brain activation to a psychological ability is logically unwarranted. Thus, McCabe and Castel's data still suggest that surplus but bogus neuro-information depicted in pictures can lead participants to draw erroneous inferences. Moreover, McCabe and Castel addressed critics' concerns by comparing subjects' responses to a brain image with their responses to an equally if not more complex multicolored topographic map of the brain, replete with all manner of scientific-looking numbers and abbreviations. The prime difference here was that the topographic brain map didn't look at all like a typical brain image. Yet participants again found the flawed explanation significantly more believable when it was accompanied by the brain image. See Martha J. Farah and Cayce J. Hook, "The Seductive Allure of 'Seductive Allure,'" *Perspectives in Biological Science*, in press but available at http://www.sas.upenn.edu/~mfarah/pdfs/The%20seductive %20allure%20of%20_seductive%20allure_%20revised.pdf. Farah and Hook report that they found "little empirical support for the claim that brain images are inordinately influential." They will publish their own data presently, personal communication with authors, December 5, 2012. Also note a failure to replicate the persuasiveness of brain images using a design like that of McCabe and Castel by David Gruber and Jacob Dickerson in their article "Persuasive Images in Popular Science: Testing Judgments of Scientific Reasoning and Credibility," *Public Understanding of Science*, in press 2013. For "brain scans show": Deena Skolnick Weisberg et al., "The Seductive Allure of Neuroscientific Explanations," *Journal of Cognitive Neuroscience* 20, no. 3 (2008): 470–477. For "brain porn" see Christopher F. Chabris and Daniel J. Simon, *The Invisible Gorilla: How Our Intuitions Deceive Us* (New York: Crown, 2010), 139.

37. For a good review of prejudice and rules of evidence, see Brown and Murphy, "Through a Scanner Darkly."

38. N. J. Schweitzer et al., "Neuroimages as Evidence in a Mens Rea Defense: No Impact," *Psychology, Public Policy, and Law* 17, no. 3 (2011): 357–393.

39. Michael Saks, personal communication to the authors, November 23, 2011. When that diagnosis of psychopathy resulted from a clinical psychiatric examination: 64.4 percent voted for death; from genetic testing, 53.4 percent; from neurological testing without presenting a neuroimage, 62.7 percent; from neurological testing with image, 46.9 percent. Also note that 61.5 percent of a control group that heard no expert testimony at all voted for death.

40. B. H. Bornstein, "The Impact of Different Types of Expert Scientific Testimony on Mock Jurors' Liability Verdicts," *Psychology, Crime, and Law* 10 (2004): 429–446; David L. Braeu and Brian Brook, "'Mock' Mock Juries: A Field Experiment on the Ecological Validity of Jury Simulation," *Law and Psychology Review* 31 (2007): 77–92; Robert M. Bray and Norbert L. Kerr, "Methodological Considerations in the Study of the Psychology of the Court," in *The Psychology of the Courtroom*, ed. Norbert L. Kerr and Robert M. Bray (New York: Academic Press, 1982): 287–323; Richard L. Wiener, Dan A. Krauss, and Joel D. Lieberman, "Mock Jury Research: Where Do We Go From Here?," *Behavioral Sciences and the Law* 29, no. 3 (2011): 467–479.

41. Sinnott-Armstrong et al., "Brain Images as Legal Evidence."

42. J. Kulynych, "Psychiatric Neuroimaging Evidence: A High-Tech Crystal Ball?," *Stanford Law Review* 49 (1997): 1249–1270; "An Overview of the Impact of Neuroscience Evidence in Criminal Law," Staff Working Paper for the President's Council on Bioethics (discussed at the council meeting in September 2004), http://bioethics.georgetown.edu/pcbe/background/neuroscience_evidence.html; Anemona Hartocollis, "In Support of Sex Attacker's Insanity Plea, a Look at His Brain," *New York Times*, May 11, 2007; *State v. Anderson*, 79 S.W.3d 420 (Mo. 2002); *People v. Kraft*, 23 Cal.4th 978 (2000). Brain explanations seem to lessen "perceived mutability" of the offender's behavior more than do psychological ones. Daniel Kahneman and Dale T. Miller, "Norm Theory: Comparing Reality to Its Alternatives," *Psychological Review* 93 (1986): 136–153.

43. John Monterosso, Edward B. Royzman, and Barry Schwartz, "Explaining Away Responsibility: Effects of Scientific Explanation on Perceived Culpability," *Ethics and Behavior* 15, no. 2 (2005): 139–158. See also Eddy Nahmias, D. Justin Coates, and Trevor Kvaran, "Free Will, Moral Responsibility, and Mechanism: Experiments in Folk Intuitions," *Midwest Studies in Philosophy* 31 (2007): 214–242; and N. J. Schweitzer and Michael J. Saks, "Neuroimage Evidence and the Insanity Defense," *Behavioral Sciences and the Law* 29, no. 4 (2011): 592–607. Jessica Gurley and David Marcus asked subjects to read about a violent crime and decide whether the defendant should be found not guilty by reason of insanity. The introduction of an image increased the insanity-acquittal rate for a psychotic defendant compared with psychiatric testimony alone, but when jurors were provided with a narrative description of how the psychotic defendant's frontal-lobe brain injury occurred (instead of an image), acquittal rates increased even more. Participants who received both the neuroimages and the brain-injury testimony rendered verdicts of not guilty by reason of insanity 47 percent of the time, whereas those who received either the neuroimages or the brain-injury testimony rendered verdicts of not guilty by reason of insanity only 31.6 percent of the time. The same general pattern occurred when the defendant was diagnosed as a psychopath. Jessica R. Gurley and David K. Marcus, "The Effects of Neuroimaging and Brain Injury on Insanity Defenses," *Behavioral Sciences and the Law* 26, no. 1 (2008): 85–97. Wendy Heath's study is reported in Wendy P. Heath et al., "Yes, I Did It, but Don't Blame Me: Perceptions of Excuse Defenses," *Journal of Psychiatry and Law* 31 (2003): 187–226. See also Dena Gromet et al., *Mind, Brain, and Character: How Neuroscience Affects People's Views of Wrongdoers* (unpublished manuscript, 2012). On sentencing by judges, see Lisa G. Aspinwall, Teneille R. Brown, and James Tabery, "The Double-Edged Sword: Does Biomechanism Increase or Decrease Judges' Sentencing of Psychopaths?" *Science* 337, no. 6096 (2012): 846–849.

44. During arguments in *Simmons*, Justice Stephen Breyer took the evidence on the biology of teen brains in stride. "I thought that the scientific evidence simply corroborated something that every parent already knows," he said. "And if it's more than that, I would like to know what more." *Roper v. Simmons*, Oral Arguments, October 13, 2004, 40, http://www.supremecourt.gov /oral_arguments/argument . . . /03-633.pdf. However, the *Simmons* ruling prompted then senator Edward Kennedy to convene a hearing in 2007 on the implications of brain science for juvenile justice. *Hearing on Adolescent Brain Development and Juvenile Justice Before the Subcommittee on Healthy Families and Communities of the Senate Committee on Education and Labor and the Subcommittee on Crime, Terrorism, and Homeland Security of the Senate Committee on the Judiciary*, 110th Cong. (July 12, 2007). One member of the children's rights litigation committee of the American Bar Association reacted to the decision as follows: "[*Simmons*] mark[s] the beginning of a legal epiphany regarding the fundamental structure and function of the adolescent brain [showing] that juveniles are less culpable for their actions than are adults"; see Hillary Harrison Gulden, "*Roper v. Simmons* and Its Application to the Daily Representation of Juveniles," *Children's Rights Litigation Committee of the ABA Section on Litigation* 7, no. 4 (2005): 3, http://apps.americanbar .org/litigation/committees/childrights/content/newsletters/childrens_fall2005.pdf. But legal scholar O. Carter Snead thinks that legal and scientific experts emphasizing neuroscience for purposes of mitigation had an impact; see O. Carter Snead, "Neuroimaging and the 'Complexity' of Capital Punishment," *New York University Law Review* 82, no. 5 (2007): 1265–1339, see specifically 1302–1308; *What Are the Implications of Adolescent Brain Development for Juvenile Justice?*, Coalition for Juvenile Justice, 2006, at http://www.issuelab.org/click/download2/applying_research_to _practice_what_are_the_implication s_of_adolescent_brain_development_for_juvenile_justice /resource_138.pdf. Simmie Baer, teleconference at the American Bar Association Center for Continuing Legal Education, "*Roper v. Simmons*: How Will This Case Change Practice in the Courtroom?" (June 22, 2005), cited in Jay D. Aronson, "Neuroscience and Juvenile Justice," *Akron Law Review* 42 (2009): 917–930, at 922. Leonard Rubenstein, executive director of Physicians for Human Rights, has stated, "Treating youth as adults goes against the foundations of juvenile justice and contradicts much of the scientific and medical knowledge around child development"; "Medical Group and Juvenile Justice Advocates Call for an End to the Incarceration of Adolescents in the Adult Criminal System," March 21, 2007, http://physiciansforhumanrights.org/press /press-releases/news-2007-03-21.html.

45. *Graham v. Florida*, 560 U.S.____, 130 S. Ct. 2011, "Brief for Petitioner" at http://www .americanbar.org/content/dam/aba/publishing/preview/publiced_preview_briefs_pdfs_07_08_08 _7412_Petitioner.authcheckdam.pdf, see 38–43; Claudia Dreifus, "Development Psychologist Says Teenagers Are Different," *New York Times*, November 3, 2009, http://www.nytimes.com/2009 /12/01/science/01conv.html, quoting developmental psychologist Laurence Steinberg, who drafted the *Simmons* brief from the American Psychological Association: "Given the fact that we know that there will be a developmental change in most people, the science says that we should give them a chance to mature out of it." The Court allowed states to impose a sentence of life without parole, however, if the sentence was imposed at an individualized sentencing proceeding. See footnote 5 in *Miller v. Alabama*, 567 U.S. _____ (2012), 9, http://www.supremecourt.gov/opinions /11pdf/10-9646g2i8.pdf; and Adam Liptak and Ethan Bronner, "Court Bars Mandatory Life Terms for Juveniles," *New York Times*, June 25, 2012, http://www.nytimes.com/2012/06/26/us /justices-bar-mandatory-life-sentences-for-juveniles.html. On the California legislation, see "Bill to Give Young Lifers a Second Chance Sent to Governor," August 20, 2012, http://sdo8.senate.ca.gov /news/2012-08-20-bill-give-young-lifers-second-chance-sent-governor. Other states are considering proposals aimed at reducing sentences for juveniles and requiring that their cases be tried in rehabilitation-friendly juvenile courts rather than in adult courts. In these instances, brain science is an explicit and much-highlighted element of lobbying efforts and policy debates. See Francis X. Shen, "Neurolegislation and Juvenile Justice" (in press).

46. In the majority opinion in *Simmons*, Justice Anthony Kennedy wrote, "As any parent knows and as the scientific and sociological studies . . . tend to confirm, '[a] lack of maturity and an underdeveloped sense of responsibility are found in youth more often than in adults.'" Opinion transcript at http://www.law.cornell.edu/supct/html/03-633.ZO.html ("543 U.S. 551"). Notably, teens' reckless behavior does not mean that teens think that they are invulnerable. Research clearly shows that they are well aware that the world can be a dangerous place. If anything, teens often undervalue the benefits of safety relative to the dangers of risk. Valerie F. Reyna and Frank Farley, "Risk and Rationality in Adolescent Decision Making: Implications for Theory, Practice, and Public Policy," *Psychological Science in the Public Interest* 7 (2006): 1–44. They are also highly sensitive to their status among peers. See also Jay N. Giedd, "The Teen Brain: Primed to Learn and Primed to Take Risks," Dana Foundation (February 26, 2009), http://www.dana.org/news/cerebrum/detail.aspx?id=19620. On awareness of the finality of death, see "Discussing Death with Children," *MedLine Plus* (most recently updated on May 2, 2011), http://www.nlm.nih.gov/medlineplus/ency/article/001909.htm; and Eva L. Essa and Colleen I. Murray, "Young Children's Understanding and Experience with Death," *Young Children* 49, no. 4 (1994): 74–81, http://webshare.northseattle.edu/fam180/topics/death/ResearchReview.htm. Experts consider the neural systems responsible for logical reasoning to be largely complete by the time young people are about sixteen, even if the circuitry involved in self-regulation continues to mature into the mid-twenties. Laurence Steinberg, "Risk Taking in Adolescence: What Changes, and Why?," *Annals of the New York Academy of Sciences* 1021 (2004): 51–58, 54.

47. Philip Graham, *The End of Adolescence* (Oxford: Oxford University Press, 2004); Robert Epstein, "The Myth of the Teen Brain," *Scientific American Mind*, April 2007, 57–63; Gene Weingarten, "Snowbound," *Washington Post Magazine*, April 26, 2005; B. J. Casey et al., "The Storm and Stress of Adolescence: Insights from Human Development and Mouse Genetics," *Developmental Psychobiology* 52, no. 3 (2010): 225–253. Perhaps, as psychologist Robert Epstein suggests, the cultural infantilizing of teens, at least in the United States, is instrumental in causing teen angst and, in turn, its corresponding brain properties. Robert Epstein, *The Case Against Adolescence: Rediscovering the Adult in Every Teen* (Fresno, CA: Quill Driver Books, 2007). Psychologist Jerome Kagan notes, "Under the right conditions, 15-year-olds can control their impulses without having fully developed frontal lobes, [otherwise] we should be having Columbine incidents every week" (referring to the 1999 high-school massacre committed by two teenage boys in Littleton, Colorado); quoted in Bruce Bower, "Teen Brains on Trial: The Science of Neural Development Tangles with the Juvenile Death Penalty," *Science News* 165, no. 19 (2004): 299–301, at 301.

48. Katherine H. Federle and Paul Skendalis, "Thinking Like a Child: Legal Implications of Recent Developments in Brain Research for Juvenile Offenders," in *Law, Mind, and Brain*, ed. Michael Freeman and Oliver R. Goodenough (Surrey, UK: Ashgate, 2009), esp. 214; "Scottsbluff Sen. John Harms [in Nebraska] said he was concerned about 18-year-olds entering into contracts and leases because of research indicating that their brains are not fully developed," in "Senators Advance Bill That Would Add Rights for Some Youth," *Unicameral Update: The Nebraska Legislature's Weekly Publication* 33, no. 4 (January 25–29, 2010), 10, http://nlc1.nlc.state.ne.us/epubs/L3000/N001-2010.pdf. On abortions, see Frederico C. de Miranda, "Parental Notification/Consent for Treatment of the Adolescent," American College of Pediatricians, Position Statement, May 17, 2010, http://www.acpeds.org/Parental-Notification/Consent-for-Treatment-of-the-Adolescent.html. See also Laurence Steinberg, "Are Adolescents Less Mature Than Adults? Minors' Access to Abortion, the Juvenile Death Penalty, and the Alleged APA 'Flip-Flop,'" *American Psychologist* 64 no. 7 (2009): 583–594; and William Saletan, "Rough Justice; Scalia Exposes a Flip Flop on the Competence of Minors," *Slate*, March 2, 2005, http://www.slate.com/id/2114219. On violent video games, see "Brief of *Amicus Curiae*: Common Sense Media in Support of Petitioners," sec. 1, July 19, 2010, http://www.americanbar.org/content/dam/aba/publishing/preview/publiced_preview_briefs_pdfs_09_10_08_1448_PetitionerAmCuCommonSenseMedia.authcheck

dam.pdf; and Jeneba Ghatt, "Supreme Court Overreaches on Video Game Ruling," *Washington Post*, June 30, 2011.

49. Francis X. Shen, "Law and Neuroscience: Possibilities for Prosecutors," *CDAA Prosecutors Brief* 33, no. 4 (2011): 17–23; O. Carter Snead, "Neuroimaging and Capital Punishment," *New Atlantis* 19 (2008): 35–63; Hughes, "Science in Court" (mentions the double-edge phenomenon in article about the Dugan case); and Brent Garland and Mark S. Frankel, "Considering Convergence: A Policy Dialogue About Behavioral Genetics, Neuroscience, and Law," *Law and Contemporary Problems* 69 (Winter/Spring 2006): 101–113. The point has also been made by others, including Nita A. Farahany and James E. Coleman Jr., "Genetics and Responsibility: To Know the Criminal from the Crime," *Law and Contemporary Problems* 69 (Winter/Spring 2006): 115–164; and Abram S. Barth, "A Double-Edged Sword: The Role of Neuroimaging in Federal Capital Sentencing," *American Journal of Law and Medicine* 33 (2007): 501–522. On danger posed by defendants, see Thomas Nadelhoffer and Walter Sinnott-Armstrong, "Neurolaw and Neuroprediction: Potential Promises and Perils," *Philosophy Compass* 7, no. 9 (2012): 631–642; and Erica Beecher-Monas and Edgar Garcia-Rill, "Danger at the Edge of Chaos: Predicting Violent Behavior in a Post-Daubert World," *Cardozo Law Review* 24 (2003): 1845–1897. Beecher-Monas and Garcia-Rill report that judges weigh a defendant's future dangerousness as heavily as, if not more heavily than, such factors as lack of remorse, mental illness, intelligence, or substance abuse. Genetic arguments can also be twisted to aggravate rather than mitigate. See Farahany and Coleman, "Genetics and Responsibility." In 2011, a federal appeals court in Manhattan overturned a six-and-a-half-year sentence in a child pornography case, saying that the judge who imposed it improperly found that the defendant would return to viewing child pornography because of an as-yet undiscovered gene. The judge, Gary L. Sharpe of the federal district court in Albany, was quoted as saying, "It is a gene you were born with. And it's not a gene you can get rid of," before he sentenced the defendant. Benjamin Weiser, "Court Rejects Judge's Assertion of a Child Pornography Gene," *New York Times*, January 28, 2011.

50. Steven K. Erickson, "The Limits of Neurolaw," *Houston Journal of Health Law and Policy* 11 (2012): 303–320, http://www.law.uh.edu/hjhlp/Issues/Vol_112/Steven%20Erickson.pdf. Farahany notes that "in many of these cases [sexually violent predator crimes], the state, and not the criminal defendant, has introduced neurological evidence to substantiate a finding of future dangerousness to justify either new civil commitment or ongoing commitment," in Nita A. Farahany, "Daily Digest," Center for Law and the Biosciences, Stanford Law School, March 16, 2011, http://blogs.law.stanford.edu/lawandbiosciences/2011/03/16/the-daily-digest-31611/. See also Fredrick E. Vars, "Rethinking the Indefinite Detention of Sex Offenders," *Connecticut Law Review* 44, no. 1 (2011): 161–195, http://uconn.lawreviewnetwork.com/files/2012/01/Vars.pdf. Adam Lamparello, "Why Wait Until the Crime Happens? Providing for the Involuntary Commitment of Dangerous Individuals Without Requiring a Showing of Mental Illness," *Seton Hall Law Review* 41, no. 3 (2011): 875–908.

51. There is a responsibility to refrain from engaging in situations, such as driving under the influence of alcohol or drugs, that set one up for considerable risk. This is illustrated by the 1956 landmark case of Emil Decina, a New York man with epilepsy who killed four children when his Buick went out of control during a seizure. Decina had been warned that he could have a seizure behind the wheel, yet he chose to drive. The court convicted him of reckless homicide because the ultimate cause was his voluntary decision to flout the risks. *People v. Decina*, 2 N.Y.2d 133 (1956). Resnick is quoted in Brian Doherty, "You Can't See Why on an fMRI: What Science Can, and Can't, Tell Us About the Insanity Defense," *Reason*, July 2007, http://reason.com/archives/2007/06/19/you-cant-see-why-on-an-fmri.

52. P. S. Applebaum, "Through a Glass Darkly: Functional Neuroimaging Evidence Enters the Courtroom," *Psychiatric Services* 60, no. 1 (2009): 21–23, 23; L. R. Tancredi and J. D. Brodie, "The Brain and Behavior: Limitations of the Legal Use of Functional Magnetic Resonance Imaging," *American Journal of Law and Medicine* 33 (2007): 271. Courts must still rely on the evidence

that mental health experts and neurologists derive from conventional tests, interviews, observations, and reports from those who know the defendant well or were present at the time of the crime. C. M. Filley, "Toward an Understanding of Violence: Neurobehavioral Aspects of Unwarranted Physical Aggression; Aspen Neurobehavioral Conference Consensus Statement," *Neuropsychiatry, Neuropsychology, and Behavioral Neurology* 14, no. 1 (2001): 1–14. The consensus statement from the Aspen Neurobehavioral Conference, written in 2000 and signed by experts in neurology, psychiatry, law, and psychology, cautioned against drawing a direct relationship between brain dysfunction and violence. "Violence occurs in a social context, and other concurrent factors such as emotional stress, poverty, crowding, alcohol and other drugs, child abuse, and disintegration of the family are involved" (3).

53. Not being able to form intent is highly unlikely except in cases in which the cognitive impairments are immense and obvious. See Laura Stephens Khoshbin and Shahram Khoshbin, "Imaging the Mind, Minding the Image: An Historical Introduction to Brain Imaging and the Law," *American Journal of Law and Medicine* 33 (2007): 171–192.

54. Ken Levy, "Dangerous Psychopaths: Criminally Responsible but Not Morally Responsible, Subject to Criminal Punishment and to Preventive Detention," *San Diego Law Review* 48 (2011): 1299.

55. For a summary of the debate, see Michael S. Gazzaniga and Megan S. Steven, "Free Will in the 21st Century: A Discussion of Neuroscience and the Law," in *Neuroscience and the Law*, ed. Brent Garland (New York: Dana Press, 2004), 52.

56. Anthony R. Cashmore, "The Lucretian Swerve: The Biological Basis of Human Behavior and the Criminal Justice System," *Proceedings of the National Academy of Sciences* 107, no. 10 (2010): 4499–4504, 4503.

Chapter 6

1. See generally Hal Higdon, *Leopold and Loeb: The Crime of the Century* (Urbana: University of Illinois Press, 1999); Simon Baatz, *For the Thrill of It: Leopold, Loeb, and the Murder That Shocked Chicago* (New York: Harper, 2008); John Theodore, *Evil Summer: Babe Leopold, Dickie Loeb, and the Kidnap-Murder of Bobby Franks* (Carbondale: Southern Illinois University Press, 2007). See also "Confession: Statement of Richard Albert Loeb," State Attorney General of Cook County, May 31, 1924, http://homicide.northwestern.edu/docs_fk/homicide/5866/LoebStatement .pdf for an in-depth description of the crime.

2. Chicago's six daily newspapers jousted for the most sensational coverage of the packed courtroom, and the trial was broadcast over radio. See "1924: Leopold and Loeb" in "Homicide in Chicago 1870–1930," at http://homicide.northwestern.edu/crimes/leopold/ for more news coverage.

3. Clarence Darrow, closing argument in *The State of Illinois v. Nathan Leopold & Richard Loeb*, delivered in Chicago, Illinois, on August 22, 1924, http://law2.umkc.edu/faculty/projects /trials/leoploeb/darrowclosing.html.

4. Clarence Darrow, *Crime: Its Cause and Treatment* (New York: Thomas Y. Cromwell, 1922), 36.

5. Judge John R. Caverly, decision and sentence in *The State of Illinois v. Nathan Leopold & Richard Loeb*, delivered in Chicago, Illinois, in 1924, http://law2.umkc.edu/faculty/projects/ftrials /leoploeb/leo_dec.htm.

6. Clarence Darrow, *Crime: Its Cause and Treatment* (New York: Thomas Y. Cromwell, 1922), 274; full text at http://www.gutenberg.org/files/12027/12027-8.txt.

7. For a useful glossary of terms and an overview, see Adina L. Roskies, "Neuroscientific Challenges to Free Will and Responsibility," *Trends in Cognitive Sciences* 10, no. 9 (2006): 419–423. On free will, see Derek Pereboom, "Living Without Free Will: The Case for Hard Incompatibilism,"

in *The Oxford Handbook of Free Will*, ed. Robert Kane (Oxford: Oxford University Press, 2002), 477–488.

8. Robert Wright, *The Moral Animal—Why We Are the Way We Are: The New Science of Evolutionary Psychology* (New York: Vintage, 1994), 338–341. Wright posits that the more we understand the genetics and evolutionary background of human nature, the more we will be drawn to utilitarian punishment policies of containment, deterrence, and rehabilitation and abandon retribution. See also David Eagleman, *Incognito: The Secret Lives of the Brain* (New York: Vintage, 2011), chap. 6; and Sam Harris, *Free Will* (New York: Free Press, 2012), 53–59, and Joshua Greene and Jonathan Cohen, "For the Law, Neuroscience Changes Everything and Nothing," *Philosophical Transactions of the Royal Society of London B: Biological Sciences* 359 (2004): 1775–1785, "new neuroscience" at 1775, "*Tout comprendre*" at 1783. Others speculate that neuroscience will actually resolve the free-will debate. See V. S. Ramachandran, *The Tell-Tale Brain: A Neuroscientist's Quest for What Makes Us Human* (New York: W. W. Norton, 2011). Ramachandran suggests that neurology can resolve questions like free will. Oliver R. Goodenough and Kristin Prehn, "A Neuroscientific Approach to Normative Judgment in Law and Justice," *Philosophical Transactions of the Royal Society of London B: Biological Sciences* 359 (2004): 1709–1726, describe how neuroscience will change conceptions of morals, justice, and normative judgment. Greene is quoted ("all behavior is mechanical") in Rowan Hooper, "Are We Puppets of Free Agents," *Wired*, Dec. 13, 2004. Ward E. Jones notes, "The origin of the French proverb '*tout comprendre, c'est tout pardonner*' ('to know all is to forgive all') is uncertain. Its earliest known appearance in exactly these words is in a Russian novel, Tolstoy's *War and Peace*." For further history, see Ward E. Jones, "Explanation and Condemnation," in *Judging and Understanding: Essays on Free Will, Narrative, Meaning and the Ethical Limits of Condemnation*, ed. Pedro Alexis Tabensky (Hampshire, UK: Ashgate Publishing, 2006), 43–44.

9. Richard Dawkins, "Let's All Stop Beating Basil's Car," January 1, 2006, http://edge.org /q2006/q06_9.html; Robert M. Sapolsky, "The Frontal Cortex and the Criminal Justice System," *Philosophical Transactions of the Royal Society of London B: Biological Sciences* 359 (2004): 1787–1796, at 1794. Darrow, closing argument in *The State of Illinois v. Nathan Leopold & Richard Loeb*.

10. Mark A. R. Kleiman, *When Brute Force Fails: How to Have Less Crime and Less Punishment* (Princeton, NJ: Princeton University Press, 2009), 88. Kleiman writes that punishment reduces crime in several ways. One is through "norm reinforcement: Changing the level of disapproval of some offense in the minds of potential offenders and those whose good opinion they value. Drunk driving and spousal abuse, having gone from being joked about or even bragged about to being widely considered disgraceful over less than a generation, provide examples. More aggressive enforcement and stiffer sentences partly result from changes in social attitudes brought about by Mothers Against Drunk Driving and the feminist movement."

11. David Eagleman, "The Brain on Trial," *Atlantic Monthly*, June/July 2011, http://www.the atlantic.com/magazine/archive/2011/07/the-brain-on-trial/308520/. On utilitarian punishment, see Richard Holton, "Introduction to Philosophy: Free Will" (class handout, University of Edinburgh, 2003), http://web.mit.edu/holton/www/edin/introfw/introfwhome.html. Holton notes, "Instead of talk of *punishing* people for anti-social behaviour, we should just think in terms of exposing them to stimuli that will make them less likely to do that thing again. In short: we subject them to treatment."

12. H. L. Mencken, *Treatise on Right and Wrong* (New York: Knopf, 1934), 88; Isaiah Berlin, "'From Hope and Fear Set Free,'" in *The Proper Study of Mankind: An Anthology of Essays* (New York: Farrar, Straus, and Giroux, 1998), 107.

13. For a superb overview of the problem of free will and determinism, see John Martin Fischer et al., *Four Views on Free Will* (Malden, MA: Blackwell Publishing, 2007); on free will, see Robert Kane, *The Oxford Handbook on Free Will* (Oxford: Oxford University Press, 2005); and Daniel

Dennett, *Elbow Room: Varieties of Free Will Worth Having* (Cambridge, MA: MIT Press, 1984). Michael Gazzaniga estimates that "98 or 99 percent" of cognitive neuroscientists share a commitment to reductive materialism in seeking to explain mental phenomena; see Jeffrey Rosen, "The Brain on the Stand," *New York Times Magazine,* March 11, 2007. In contrast, philosophers are more heterogeneous, according to a survey titled "Free Will: Compatibilism, Libertarianism, or No Free Will?" that was conducted by David Chalmers in November 2009, results at http://philpapers .org/surveys/Survey. The survey was taken by 931 select analytic philosophers. Out of 931 respondents, 59 percent accepted or leaned toward compatibilism; 13.7 percent accepted or leaned toward libertarianism; 12.3 percent accepted or leaned toward no free will; and 14.9 percent were "other." Jerry Coyne is quoted in "You Don't Have Free Will," *The Chronicle Review,* March 18, 2012, http://chronicle.com/article/Jerry-A-Coyne/131165/. Determinism is also impossible to prove, as Coyne notes. Maybe you had only one choice, but it seems impossible to prove that you could not have done something else; cited in Harris, *Free Will,* 76n17.

14. "Causal vacuum" is a term coined by philosopher Patricia Churchland. Patricia Churchland, "The Big Questions: Do We Have Free Will?," New Scientist, November 2006, http://philosophy faculty.ucsd.edu/faculty/pschurchland/papers/newscientist06dowehavefreewill.pdf. Some thinkers have invoked quantum mechanics as a way to refute indeterminism. We are, after all, made of electrons, and they do not follow Newtonian physics. Electrons inhabit a cloud, a distribution of many different states at the same time. There is debate, however, over whether random subatomic events (quantum indeterminacy) affect brain function and human behavior. Some people say yes; others say no. See Roskies, "Neuroscientific Challenges to Free Will and Responsibility."

15. Nancey Murphy and Warren Brown, *Did My Neurons Make Me Do It? Philosophical and Neurobiological Perspectives on Free Will* (Oxford: Oxford University Press, 2009); see especially chapter 5.

16. "Hume on Free Will," *Stanford Encyclopedia of Philosophy,* "1. Two Kinds of 'Liberty': The Basics of the Classical Interpretations," December 14, 2007, http://plato.stanford.edu/entries /hume-freewill/. Janet Radcliffe Richards, *Human Nature After Darwin: A Philosophical Introduction* (London: Routledge, 2000), 148. Also, according to philosopher Harry Frankfurt, the freedom to do otherwise is not the kind of freedom required for moral responsibility; Harry Frankfurt, *The Importance of What We Care About* (Cambridge: Cambridge University Press, 1998), viii. Note that philosophers debate the meaning and intelligibility of the notion "to be able to do otherwise," also called the principle of alternative possibilities. For a reader-friendly overview, see "Could Have Done Otherwise," *The Information Philosopher,* http://www.informationphilosopher .com/freedom/otherwise.html. See also Ronald Bailey, "Pulling Our Own Strings," *Reason,* May 2003, 24–31; "The difference between a responsible brain and a nonresponsible one," says philosopher Daniel Dennett, "[is] a difference in the capacity of that brain to respond to information, to respond to reason, to be able to reflect."

17. Roy F. Baumeister, William A. Crescioni, and Jessica L. Alquist, "Free Will as Advanced Action Control for Human Social Life and Culture," *Neuroethics* 4, no. 1 (2011): 1–11; Eddy Nahmias, "Why 'Willusionism' Leads to 'Bad Results': Comments on Baumeister, Crescioni, and Alquist," *Neuroethics* 4, no. 1 (2011): 17–24; Shaun Nichols, "Experimental Philosophy and the Problem of Free Will," *Science* 331 (2011): 1401–1403 ("compatibilist views seem to be favored when emotional attitudes are triggered" [1403]); Eddy Nahmias et al., "Is Incompatibilism Intuitive?," *Philosophy and Phenomenological Research* 73 (2006): 28–53; Shaun Nichols and Joshua Knobe, "Moral Responsibility and Determinism: The Cognitive Science of Folk Intuitions," *Nous* 43 (2007): 663–685; Adina Roskies and Shaun Nichols, "Distance, Anger, Freedom: An Account of the Role of Abstraction in Compatibilist and Incompatibilist Intuitions," *Philosophical Psychology* 24, no. 6 (2011): 803–823; Tamler Sommers, "Experimental Philosophy and Free Will," *Philosophy Compass* 5, no. 2 (2010): 199–212; and "Critiques of xphi," *Experimental Philosophy* (blog), http://pantheon.yale.edu/~jk762/xphipage/Experimental%20Philosophy-Critiques.html. Vohs and

Schooler had subjects read one of three passages: the first encouraged a belief in determinism ("Ultimately, we are biological computers—designed by evolution, built through genetics, and programmed by the environment"), the second asserted free will ("I am able to override the genetic and environmental factors that sometimes influence my behavior"), and the third was a neutral passage on agriculture. Subjects who read the anti-free-will passages displayed a greater tendency to cheat on a problem-solving task than did those who read the other passages; Kathleen D. Vohs and Jonathan W. Schooler, "The Value of Believing in Free Will: Encouraging a Belief in Determinism Increases Cheating," *Psychological Science* 19 (2008): 49–54. In other versions of this experiment, subjects primed to believe in determinism were less willing to help others, more inclined to goof off on the job, and, in one notable instance, more apt to add hot sauce to a dish so as to offend people with a known aversion to spicy food. For the results on goofing off, see Tyler F. Stillman et al., "Personal Philosophy and Personal Achievement: Belief in Free Will Predicts Better Job Performance," *Social Psychological and Personality Science* 1, no. 1 (2010): 43–50. For the results on hot sauce, see Roy F. Baumeister, E. J. Masicampo, and C. N. DeWall, "Prosocial Benefits of Feeling Free: Disbelief in Free Will Increases Aggression and Reduces Helpfulness," *Personality and Social Psychology Bulletin* 35, no. 2 (2009): 260–268. Along with behaving badly as a result of seeing things through the lens of determinism, people may also adopt a less punitive stance. Psychologist Azim Shariff and colleagues found that when subjects were primed with anti-free-will messages, they reported less punitive attitudes toward hypothetical murderers and greater forgiveness of interpersonal transactions, consistent with the notion that actors may regard both themselves and others as less accountable in a deterministic milieu. Azim Shariff et al., "Diminished Belief in Free Will Increases Forgiveness and Reduces Retributive Punishment," *Psychological Sciences*, in preparation. One crime was designed to arouse strong emotions (Bill stalking and raping a woman); another was much less provocative (Mark cheating on his taxes). Two-thirds of subjects said that Bill was fully responsible for his actions, but just 23 percent held Mark fully accountable. Working with a team of international researchers, Knobe and Nichols replicated these findings in a range of countries, including the United States, Hong Kong, India, and Colombia; see H. Sarkissian et al., "Is Belief in Free Will a Cultural Universal?" *Mind and Language* 25, no. 2 (2010): 346–358. In the end, philosophers and psychologists debate whether the scenarios painted suggest to subjects a determined world where conscious mental states direct actors' behaviors or a world in which conscious mental states are completely bypassed. Depending on subjects' interpretation, their responses will most likely vary. Apart from this significant methodological issue, the pressing question is under what circumstances (e.g., situational, cultural, characterological, emotional, environmental) people have different intuitions about responsibility, and we must keep in mind that our moral sensibilities deepen and transform as we experience the world. See A. Feltz and E. T. Cokely, "Do Judgments About Freedom and Responsibility Depend on Who You Are? Personality Differences in Intuitions About Compatibilism and Incompatibilism," *Consciousness and Cognition* 18, no. 1 (March 2009): 342–350; and David A. Pizarro and Erik G. Helzer, "Freedom of the Will and Stubborn Moralism," in *Free Will and Consciousness: How Might They Work?*, ed. Roy F. Baumeister, Alfred R. Mele, and Kathleen D. Vohs (Oxford: Oxford University Press, 2010), 101–120 (showing how our cognitive bias works in the direction of ascribing greater agency to an actor when that actor's deeds are bad than when they are good).

18. Note that determinism and illusionism are overlapping but by no means identical perspectives. Both categorically reject the notion that people choose their actions, and, therefore, both deny the concept of "free will." But although all illusionists are determinists, not all determinists are illusionists. That is, not all determinists share the belief that conscious states are enlisted only to help us justify our actions after the fact.

19. Benjamin Libet et al., "Time of Conscious Intention to Act in Relation to Onset of Cerebral Activity (Readiness-Potential): The Unconscious Initiation of a Freely Voluntary Act," *Brain* 106 (1983): 623–642; John-Dylan Haynes, "Decoding and Predicting Intentions," *Annals of the New*

York Academy of Sciences 1224, no. 1 (2011): 9–21. In an experiment analogous to Libet's but using fMRI, Haynes could predict with a middling 60 percent accuracy whether a person would use his right or left hand to press a button a full seven seconds before the person was aware of deciding which one he would push.

20. Sukhvinder S. Obhi and Patrick Haggard, "Free Will and Free Won't," *American Scientist*, July–August 2004, 358–365, http://www.americanscientist.org/template/AssetDetail/assetid/34008/page/5.

21. Benjamin Libet, *Mind Time: The Temporal Factor in Consciousness* (Cambridge, MA: Harvard University Press, 2004), 137–138. See generally Daniel M. Wegner, *The Illusion of Conscious Will* (Cambridge, MA: MIT Press, 2002); and John Tierney, "Is Free Will Free?," *New York Times*, June 19, 2006, http://tierneylab.blogs.nytimes.com/2009/06/19/is-free-will-free/.

22. Emily Pronin et al., "Everyday Magical Powers: The Role of Apparent Mental Causation in the Overestimation of Personal Influence," *Journal of Personal and Social Psychology* 91, no. 2 (2006): 218–231.

23. Wegner, *The Illusion of Conscious Will*; Wegner also explores kinds of automatic situations in which individuals' actions seem to happen to them rather than being caused by them, such as hypnosis-induced behavior, Ouija-board spelling, automatic writing, and trance channeling, all of which are caused by the actor. These situations illustrate that the feeling of conscious will does not always correlate with the true causes of action. Michael S. Gazzaniga, *Who's in Charge?: Free Will and the Science of the Brain* (New York: Harper Collins, 2011), 82–89. Gazzaniga gives a lengthy description of confabulation in "split-brain" patients whose epilepsy was treated surgically by severing the band of nerve fibers connecting the left and right hemispheres. Gazzaniga discovered an operation performed in the left hemisphere that enables us to interpret our behavior and responses, cognitive or emotional, to environmental stimuli. "The interpreter," he says, constantly establishes a running narrative of our actions, emotions, thoughts, and dreams. It is the glue that keeps our story unified and creates our sense of being a coherent, rational agent. Confabulation can be observed in people with stroke, with other neurological conditions, and under hypnosis.

24. Timothy Wilson, *Strangers to Ourselves: Discovering the Adaptive Unconscious* (Cambridge, MA: Harvard University Press, 2004).

25. For a review, see Roy F. Baumeister, E. J. Masicampo, and Kathleen D. Vohs, "Do Conscious Thoughts Cause Behavior?," *Annual Review of Psychology* 62 (2011): 331–361.

26. Determinism, after all, does not relieve us of the need to make decisions, as philosopher Hilary Bok points out in "Want to Understand Free Will? Don't Look to Neuroscience," *The Chronicle Review*, March 18, 2012, http://chronicle.com/article/Hilary-Bok/131168/.

27. Roy F. Baumeister points out, "It is ironic that many researchers who claim to demonstrate the dispensability of conscious thought give their study subjects crucial instructions that must be processed consciously. In doing so, researchers are relying heavily on the very faculty they believe they have discredited." Personal communication with authors, June 6, 2012. Also, consider the practical benefits of believing that one controls one's outcomes, which is critical to changing behavior. Volumes of data show that people with a greater sense of self-efficacy are much more likely to succeed in quitting smoking or losing weight than those who don't. For self-efficacy, see generally Albert Bandura, *Self-efficacy: The Exercise of Control* (New York: Freeman, 1997).

28. Will Durant, *The Story of Philosophy* (New York: Pocket Books, 1991), 76.

29. Shaun Nichols, "The Folk Psychology of Free Will: Fits and Starts," *Mind and Language* 19 (2004): 473–502.

30. Nichols and Knobe, "Moral Responsibility and Determinism"; Hagop Sarkissian et al., "Is Belief in Free Will a Cultural Universal?," *Mind and Language* 25, no. 3 (2010): 346–358. Authors reported that Chinese, Indian, and Colombian populations rejected that human decision making is such that people do not choose. Nadia Chernyak et al., "A Comparison of Nepalese and American Children's Concepts of Free Will," *Proceedings of the 33rd Annual Meeting of the Cognitive Sci-*

ence Society, ed. Laura Carlson, Christoph Hoelscher, and Thomas F. Shipley (Austin, TX: Cognitive Science Society, 2011), 144–149.

31. "The violation of justice is injury: it does real and positive hurt to some particular persons, from motives which are naturally disapproved of. It is, therefore, the proper object of resentment, and of punishment, which is the natural consequence of resentment." Adam Smith, *The Theory of Moral Sentiments*, chap. 1, sec. II, pt. II (London: A. Millar, 1790), Library of Economics and Liberty (online), http://www.econlib.org/library/Smith/smMS2.html. See also Paul H. Robinson and Robert Kurzban, "Concordance and Conflict in Intuitions of Justice," *Minnesota Law Review* 91, no. 6 (2007): 1829–1907; David A. Pizarro and E. K. Helzer, "Stubborn Moralism and Freedom of the Will," in *Free Will and Consciousness: How They Might Work*, ed. Roy F. Baumeister, Alfred R. Mele, and Kathleen D. Vohs (Oxford: Oxford University Press, 2011), 101–120. Some social animals manifest rudimentary perceptions of fairness as well. See Megan van Wolkenton, Sarah F. Brosnan, and Frans B. M. de Waal, "Inequity Responses of Monkeys Modified by Effort," *Proceedings of the National Academy of Sciences* 104, no. 47 (2007): 18854–18859. The authors found that capuchin monkeys seemed to register displeasure when a fellow monkey got a better reward (a nice fat grape compared with a boring piece of cucumber) for performing the same task. The cucumber recipients hurled their reward at the experimenter. See also Friederike Range et al., "The Absence of Reward Induces Inequity Aversion in Dogs," *Proceedings of the National Academy of Sciences* 106, no. 1 (2009): 340–345; and Leda Cosmides and John Tooby, "Neurocognitive Adaptations Designed for Social Exchange," in *Handbook of Evolutionary Psychology*, ed. David M. Buss (Hoboken, NJ: John Wiley and Sons, 2005), 584–627; On detecting and punishing cheaters, see Elsa Ermer, Leda Cosmides, and John Tooby, "Cheater-Detection Mechanisms," in *Encyclopedia of Social Psychology*, ed. Roy F. Baumeister and Kathleen D. Vohs (Thousand Oaks, CA: Sage, 2007), 138–140 (speculating that moral emotions, such as pride and shame, likely evolved alongside cooperative human behavior because ancestors who lived within groups of individuals motivated by these sentiments were more likely to survive [139]); see also Alan P. Fiske, *Structures of Social Life* (New York: Free Press, 1991); Shalom H. Schwartz and Wolfgang Bilsky, "Toward a Theory of the Universal Content and Structure of Values: Extensions and Cross-Cultural Replications," *Journal of Personality and Social Psychology* 58 (1990): 878–891; Richard A. Shweder et al., "The 'Big Three' of Morality (Autonomy, Community, and Divinity) and the 'Big Three' Explanations of Suffering," in *Morality and Health*, ed. Allan M. Brandt and Paul Rozin (London: Routledge, 1997), 119–169; and Morris B. Hoffman and Timothy H. Goldsmith, "The Biological Roots of Punishment," *Ohio State Journal of Criminal Law* 1 (2004): 627–641. See generally Samuel Bowles and Herbert Gintis, *A Cooperative Species: Human Reciprocity and Its Evolution* (Princeton, NJ: Princeton University Press, 2012); Alan Fiske, "Four Elementary Forms of Sociality: Framework for a Unified Theory of Social Relations," *Psychological Review* 99, no. 4 (1992): 689–732; and Donald E. Brown, *Human Universals* (New York: McGraw-Hill Humanities, 1991).

32. Jonathan Haidt and Craig Joseph, "Intuitive Ethics: How Innately Prepared Intuitions Generate Culturally Variable Virtues," *Daedalus: On Human Nature* 133, no. 4 (2004): 55–66. See also John Mikhail, "Universal Moral Grammar: Theory, Evidence, and the Future," *Trends in Cognitive Sciences* 11, no. 4 (2007): 143–152. The quotation is from Haidt and Joseph, "Intuitive Ethics," 55. This is not to say, however, that all facets of moral behavior are emphasized equally across all cultures. Joseph Henrich et al., "Markets, Religion, Community Size, and the Evolution of Fairness and Punishment," *Science* 327 (2010): 1480–1484, looked at fifteen diverse populations and found that people's propensities to behave kindly to strangers and to punish unfairness are strongest in large-scale communities with market economies, where such norms are essential to the smooth functioning of trade. "These results suggest that modern prosociality is not solely the product of an innate psychology, but also reflects norms and institutions that have emerged over the course of human history" (1480).

33. Daniel Kahneman, Jack L. Knetsch, and Richard H. Thaler, "Fairness and the Assumptions of Economics," *Journal of Business* 59, no. 4 (1986): S285–S300.

34. Even younger children show an affinity for benign actors, as shown in the behavior of eight-month-old babies who preferred—that is, reached for—the "nice" stuffed animals that behaved kindly toward other stuffed critters over "bad" ones who thwarted them. Paul Bloom, "Moral Nativism and Moral Psychology," in *The Social Psychology of Morality: Exploring the Causes of Good and Evil*, ed. Mario Mikulincer and Phillip R. Shaver (Washington, DC: American Psychological Association, 2012), 71–89; also Stephanie Sloane, Renee Baillargeon, and David Premack, "Do Infants Have a Sense of Fairness?," *Psychological Science* 23, no. 2 (2012): 196–204; and Judith Smetana et al., "Developmental Changes and Individual Differences in Young Children's Moral Judgments," *Child Development* 83, no. 2 (2012): 683–696.

35. Philip E. Tetlock, William T. Self, and Ramadhar Singh, "The Punitiveness Paradox—When Is External Pressure Exculpatory and When a Signal Just to Spread Blame?," *Journal of Experimental and Social Psychology* 46, no. 2 (2010): 388–395.

36. Cited in Jonathan Haidt and John Sabini, "What Exactly Makes Revenge Sweet?" (unpublished manuscript, University of Virginia, 2004).

37. Kevin Carlsmith and John M. Darley, "Psychological Aspects of Retributive Justice," *Advances in Experimental Social Psychology* 40 (2008): 199, 207.

38. On the practical value of censure, see Harris, *Free Will*, 56. See also Sarah Mathew and Robert Boyd, "Punishment Sustains Large-Scale Cooperation in Pre-state Warfare," *Proceedings of the National Academy of Sciences* 108, no. 28 (2011): 11375–11380; Benedikt Herrmann, Christian Thöni, and Simon Gächter, "Antisocial Punishment Across Societies," *Science* 319 (2008): 1362–1367; Robert Boyd, Herbert Gintis, and Samuel Bowles, "Coordinated Punishment of Defectors Sustains Cooperation and Can Proliferate When Rare," *Science* 328 (2010): 617–620.

39. Note that social control and reduction of future crime may be welcome by-products of retributive punishment but are beside the point.

40. The so-called expressive function of punishment is well described by Jean Hampton, "The Moral Education Theory of Punishment," *Philosophy and Public Affairs* 13, no. 3 (1984): 208, 215–217, 227; and Joel Feinberg, "The Expressive Function of Punishment," in *Doing and Deserving* (Princeton, NJ: Princeton University Press, 1970), 95–101. In a classic treatment, the nineteenth-century French sociologist Emile Durkheim claimed that community enforcement of laws strengthened bonds of social solidarity. See Emile Durkheim, *The Division of Labor in Society* (New York: Free Press, 1997), 34–41. The Castrex example is inspired by James Q. Wilson, "The Future of Blame," *National Affairs*, Winter 2010, 105–114. There are serious strains of *A Clockwork Orange* here. In the 1971 film of the same name, based on the 1962 dystopian novel by Anthony Burgess, a violent hoodlum/rapist is subjected to an aversion technique in which he is shown violent films while he is given a drug that makes him nauseated. The sole goal is to prevent future violence. Within two weeks, the mere thought of violence or sex makes him ill. The minister of the interior pronounces him "cured," but the prison chaplain differs, saying that "there's no morality without choice." From the standpoint of the observing community, what if someone else now says, "Ah, I can rape and get away with it in that I'll only have to spend a few weeks in prison and take a pill"? This dire consequence is taken into consideration under the umbrella of the deterrence function of the law—which opponents of retribution do support—and alone would justify a more arduous sentence for the rapist. On retribution as a mechanism for restoring equality between wrongdoers and law-abiding citizens, see John Finnis, "Retribution: Punishment's Formative Aim," *American Journal of Jurisprudence* 44 (1999): 91–103. Kenworthey Bilz notes, "For one thing, research consistently shows that victims and third parties alike are motivated to punish not out of instrumental motives such as deterrence, incapacitation, or rehabilitation, but out of a desire for retribution." Kenworthey Bilz, "The Puzzle of Delegated Revenge," *Boston University Law Review* 87 (2007): 1088. A study conducted by the late psychology professor Kevin M. Carl-

smith found that "when people sentence criminals, they do so from a retributive rather than utilitarian stance." Kevin M. Carlsmith, "The Roles of Retribution and Utility in Determining Punishment," *Journal of Experimental Social Psychology* 42 (2006): 446. See also Kevin M. Carlsmith, John M. Darley, and Paul H. Robinson, "Why Do We Punish? Deterrence and Just Deserts as Motives for Punishment," *Journal of Personality and Social Psychology* 83 (2002): 284–299. Carlsmith, Darley, and Robinson lend support to the idea that people favor a "just-deserts" theory (in which punishers are concerned with "providing punishment appropriate to the given harm") over a theory of deterrence (which "holds that an offender's punishment should be just sufficient to prevent future instances of the offence") in punishing wrongdoers. See also Kevin M. Carlsmith, John M. Darley, and Paul H. Robinson, "Incapacitation and Just Deserts as Motives for Punishment," *Law and Human Behavior* 24, no. 6 (2000): 659, 676. In the *New Yorker*, Jared Diamond writes of his late father-in-law, whose mother, sister, and niece were killed in the Holocaust. The survivor had the chance to kill the gang member responsible but decided to turn him over to the police. The killer was imprisoned for only a year. Diamond's father-in-law was consumed for the rest of his life by regret and guilt for having inadvertently allowed his family's murderer to go free. Jared Diamond, "Annals of Anthropology: Vengeance Is Ours," *New Yorker*, April 21, 2008, 74–89, http://www.unl.edu/rhames/courses/war/diamond-vengeance.pdf. Samuel R. Gross and Phoebe C. Ellsworth, "Hardening of the Attitudes: Americans' Views on the Death Penalty," *Journal of Social Issues* 50, no. 2 (1994): 27–29, find that Americans most often cite retribution as a rationale for supporting the death penalty. On motivations for punishment, see Peter French, *The Virtues of Vengeance* (Lawrence: University Press of Kansas, 2001); Jeffrie G. Murphy, "Two Cheers for Vindictiveness," *Punishment and Society* 2, no. 2 (2000): 131–143; and generally, William Ian Miller, *Eye for an Eye* (Cambridge, UK: Cambridge University Press, 2006).

41. Clarence Darrow, *The Story of My Life* (New York: Da Capo, 1996), "hanged" at 238, "abusive" letters at 233. Suzan Clarke, "Casey Anthony Verdict: Anthony Family Gets Death Threats in Wake of Acquittal, Asks for Privacy," *ABC News*, July 5, 2011, http://abcnews.go.com/US/casey-anthony-verdict-anthony-family-death-threats-wake/story?id=14004306#.UJhDjHglZFI; Benjamin Weiser, "Judge Explains 150-Year Sentence for Madoff," *New York Times*, June 29, 2011, http://www.nytimes.com/2011/06/29/nyregion/judge-denny-chin-recounts-his-thoughts-in-bernard-madoff-sentencing.html?hp#.

42. Donald Black, *The Behavior of the Law, Special Edition* (Bingley, UK: Emerald Group Publishing, 2010). It is a troubling enough truth of our system—and surely other societies—that the severity of penalties inflicted on wrongdoers fluctuates with the social status of the victim. "If the offender is more educated than his victim, the seriousness of the offense decreases" (66). Paul H. Robinson and John Darley, "The Utility of Desert," *Northwestern University Law Review* 91, no. 2 (1997): 458–497, conclude that the "traditional utilitarian theories of deterrence, incapacitation, and rehabilitation . . . have little effect in many instances [and] instead that the real power to gain compliance with society's rules of prescribed conduct lies not in the power of the intertwined forces of social and individual moral control. . . . Criminal law, in particular, plays a central role in creating and maintaining the social consensus necessary for sustaining moral norms" (458). See also Jean Hampton, "An Expressive Theory of Retribution," in *Retributivism and Its Critics*, ed. Wesley Cragg (Stuttgart, Ger.: Franz Steiner Verlag, 1992), 5 ("The crime represents the victim as demeaned relative to the wrongdoer; the punishment 'takes back' the demeaning message"). In a series of experiments, Jennifer Kenworthey Bilz found that being victimized tends to reduce the victim's standing in her own eyes and in the eyes of those around her, and that the effect is worse if no retribution is exacted from the perpetrator. Jennifer Kenworthey Bilz, "The Effect of Crime and Punishment on Social Standing" (Ph.D. diss., Princeton University, 2006), 72–73. Unfortunately, her sample size was limited to twenty undergrads. See also Kenworthey Bilz and John M. Darley, "What's Wrong with Harmless Theories of Punishment?," *Chicago-Kent Law Review* 79 (2004): 1215–1252. Failure to punish offenders raised their social status only a little, though, relative to the baseline. Bilz speculates that maybe third

parties read more into the meaning of punishment for the victim than for the offender but does not know why. See Bilz, "Effect of Crime and Punishment on Social Standing," 42, fig. 2.

43. Melvin J. Lerner, *The Belief in a Just World: A Fundamental Delusion* (New York: Plenum Press, 1980).

44. Melvin J. Lerner and Dale T. Miller, "Just World Research and the Attribution Process: Looking Back and Ahead," *Psychological Bulletin* 85, no. 5 (1978): 1030–1051, 1032; see 1050–1051 for more studies confirming this observation/interpretation. See also A. Lincoln and George Levinger, "Observers' Evaluations of the Victim and the Attacker in an Aggressive Incident," *Journal of Personality and Social Psychology* 22, no. 2 (1972): 202–210. In this study, researchers presented subjects with the report of an innocent victim of a policeman's attack. Subjects' ratings of the victim were more negative when they (the subjects) could not lodge the complaint against the policeman than when they could. In a study by Robert M. McFatter, "Sentencing Strategies and Justice: Effects of Punishment Philosophy on Sentencing Decisions," *Journal of Personality and Social Psychology* 36, no. 12 (1978): 1490–1500, subjects who imposed the least severe penalties on the offender blamed the victim of the crime more than did those who punished the offender more harshly.

45. Tom R. Tyler, *Why People Obey the Law* (Princeton, NJ: Princeton University Press, 2006), 19–69. Cathleen Decker, "Faith in Justice System Drops," *Los Angeles Times*, October 8, 1995, S2; see also Cathleen Decker and Sheryl Stolberg, "Half of Americans Disagree with Verdict," *Los Angeles Times*, October 4, 1995, A1 (reporting a similar lack of confidence in the criminal justice system in a national poll not limited to Los Angeles residents); and Alexander Peters, "Poll Shows Courts Rate Low in Public Opinion," *Recorder*, December 11, 1992, 1. Lawful behavior is sustained more powerfully by fear of social disapproval than by threat of formal legal sanction, according to Harold G. Grasmick and Robert Bursick, "Conscience, Significant Others, and Rational Choice: Extending the Deterrence Model," *Law and Society Review* 24 (1990): 837–861, esp. 854. When people deem legal authority legitimate, they feel that they ought to defer to official decisions and rules, following them out of civic obligation rather than out of fear of punishment or anticipation of reward, according to Tom Tyler, "Psychological Perspectives on Legitimacy and Legitimation," *Annual Review of Psychology* 57 (2006): 375–400. On witness behavior, see Kevin M. Carlsmith and John M. Darley, "Psychological Aspects of Retributivist Justice," *Advances in Experimental Social Psychology* 40 (2008): 193–263. Legal scholar Janice Nadler examined subjects' reactions to the true case of eighteen-year-old David Cash. In 1997, Cash and a friend visited a Nevada casino, where Cash watched his friend restrain and fondle a seven-year-old girl in the casino bathroom and walked out shortly before the friend raped and murdered the girl. When the friend met up with Cash later, he told him what he had done. The young men spent the next two days gambling. Nadler gave subjects two scenarios. In the "just outcome," Cash is prosecuted for being an accessory to murder after the fact and spends a year in prison. In the "unjust outcome" (which happened to be the true disposition of the case), Cash goes free. Subjects in the "just-outcome" group were more likely to comply with a judge's instructions in a subsequent case of theft, while participants primed with the unjust outcome (Cash goes free) exhibited a greater rate of noncompliance. Janice Nadler, "Flouting the Law," *Texas Law Review* 83 (2005): 1339–1441, 1423–1424 (description of the experiment). On jury nullification, see James M. Keneally, "Jury Nullification, Race, and *The Wire*," *New York Law School Law Review* 55 (2010–2011): 941–960, http://www.nyls.edu/user_files/1/3/4/17/49/1156/Law%20Review%2055.4_01Keneally.pdf.

46. See generally Susan Herman, *Parallel Justice for Victims of Crime* (self-published, 2010), http://www.paralleljustice.org/thebook/. Herman notes that when victims' intuitions of justice are satisfied, their belief in a just world is supported. See also Lawrence W. Sherman and Heather Strang, "Repair or Revenge: Victims and Restorative Justice," *Utah Law Review* 15, no. 1 (2003): 1–42; and Melvin J. Lerner and Leo Montada, eds., *Responses to Victimizations and Belief in a Just World* (New York: Plenum Press, 1998). Lerner and Montada's review of research speaks to the value that a belief in a just world can have for victims, supporting the idea that victims need legal

acknowledgment that the crimes against them were wrong and morally offensive. As they explain in their preface, "A rather new line of research describes the functions of BJW [belief in a just world] in victims' coping with their own hardships and problems. Several contributors report evidence that BJW helps to protect people against a stressful negative view of their situation, especially their fears of being unjustly victimized. In this manner, BJW seems to function as a resource for victims, too" (viii). Note that although there is extensive argumentation in the international legal community for postwar tribunals and truth and reconciliation commissions (TRCs), there is exceedingly little quantitative data regarding their efficacy in a cathartic sense for victims. See generally Michal Ben-Josef Hirsch, Megan MacKenzie, and Mohamed Sesay, "Measuring the Impacts of Truth and Reconciliation Commissions: Placing the Global 'Success' of TRCs in Local Perspective," *Cooperation and Conflict* 47, no. 3 (2012): 386–403; and Neil J. Kritz, "Coming to Terms with Atrocities: A Review of Accountability Mechanisms for Mass Violations of Human Rights," *Law and Contemporary Problems* 59, no. 4 (1996): 127–128. Kritz notes that truth commissions add "a meaningful acknowledgment of past abuses by an official body perceived domestically and internationally as legitimate and impartial. Such an entity cannot substitute for prosecutions—and rarely affords those implicated in their inquiry the due process protections to which they are entitled in a judicial proceeding—but it can serve many of the same purposes, to the extent that it (1) provides the mandate and authority for an official investigation of past abuses; (2) permits a cathartic public airing of the evil and pain that has been inflicted, resulting in an official record of the truth; (3) provides a forum for victims and their relatives to tell their story, have it made part of the official record, and thereby provide a degree of societal acknowledgment of their loss; and (4) in some cases, establishes a formal basis for subsequent compensation of victims and/or punishment of perpetrators." In discussing the International Criminal Tribunal for the Former Yugoslavia (ICTY), Payam Akhavan writes, "Truth telling will also enable victims to hear and see their stories told—either their own personal stories or stories like theirs—in an officially sanctioned forum before the international community. . . . Ambassador Albright explained before the Security Council that 'among the millions' who will learn of the ICTY's establishment 'are the hundreds of thousands of civilians who are the victims of horrific war crimes against humanity in the former Yugoslavia. To these victims we declare by this action that your agony, your sacrifice, and your hope for justice have not been forgotten.' In this connection, it is important to emphasize that recollection and recognition of the past is a highly valuable commodity for victims that is often underestimated by external observers." Payam Akhavan, "Justice in the Hague, Peace in the Former Yugoslavia? A Commentary on the United Nations War Crimes Tribunal," *Human Rights Quarterly* 20, no. 4 (1998): 766–767.

47. Roskies, "Neuroscientific Challenges to Free Will and Responsibility," elaborates the arguments for uncoupling free will from legal and moral ideas of responsibility. See also Dennett, *Elbow Room*. Morse's comment is from Stephen J. Morse, "The Non-problem of Free Will in Forensic Psychiatry and Psychology," *Behavioral Sciences and the Law* 25 (2007): 203–220.

48. Closing argument, *The State of Illinois v. Nathan Leopold and Richard Loeb*, delivered by Clarence Darrow, Chicago, Illinois, August 22, 1924, at http://law2.umkc.edu/faculty/projects/ftri als/leoploeb/darrowclosing.html.

49. In his closing argument, Darrow said, referring to Loeb, "I have never in my life been so interested in fixing blame as I have been in relieving people from blame. . . . It would be the height of cruelty [to] visit the [death] penalty upon this poor boy." Ibid. "The law deals firmly but mercifully with individuals whose behavior is obviously the product of forces beyond their control," Greene and Cohen say, referring to severely mentally ill offenders. "Someday the law may treat all convicted criminals this way. That is, humanely." Greene and Cohen, "For the Law, Neuroscience Changes Everything and Nothing," 1783. Sam Harris writes, "This shift in understanding [the causes of human behavior] represents progress toward a deeper, more consistent, and more compassionate review of our common humanity" (*Free Will*, 55). Luis E. Chiesa, "Punishing Without Free Will," *Utah Law Review*, no. 4 (2011): 1403–1460, argues against blame in the name of

efficiency and humanity. Nick Trakaskis, "Whither Morality in a Hard Determinist World?," *Sorites* 14 (2007): 14–40, http://www.sorites.org/Issue_19/trakakis.htm, argues that determinism should reduce rage and vengeance and lead to more altruism and empathy. Kelly Burns and Antoine Bechara, "Decision Making and Free Will: A Neuroscience Perspective," *Behavioral Sciences and the Law* 25, no. 2 (2007): 263–280, write of neuroscience as undermining the legal conception of freedom of the will and thus introducing a more humane and effective criminal justice system.

50. Esteemed British philosopher Peter F. Strawson asserts that we express attitudes when we hold people accountable; those attitudes embrace a range of attitudes deriving from our participation in personal relationships, e.g., resentment, indignation, hurt feelings, anger, gratitude, reciprocal love, and forgiveness. P. F. Strawson, ed., *Freedom and Resentment and Other Essays* (New York: Routledge, 2008), 5.

51. Strawson argues that our ordinary practices involve us in the reactive attitudes, attitudes that treat people as participants in a community rather than objects to be studied, attitudes like resentment, indignation, gratitude, reciprocal love, forgiveness, and obligation; these attitudes and practices are so basic and fundamental that no theory of free will could change them, whatever the deep metaphysical truth on these issues (*Freedom and Resentment*, 1–28). Tamler Sommers, *Relative Justice: Cultural Diversity, Free Will, and Moral Responsibility* (Princeton, NJ: Princeton University Press, 2012), 173–202.

Epilogue

1. Neuroskeptic, "fMRI Reveals True Nature of Hatred," *Neuroskeptic* (blog), October 30, 2008, http://neuroskeptic.blogspot.com/2008/10/fmri-reveals-true-nature-of-hatred.html.

2. Jeffrey Rosen, "The Brain on the Stand," *New York Times Magazine*, March 11, 2007.

3. Apoorva Mandavilli, "Actions Speak Louder Than Images—Scientists Warn Against Using Brain Scans for Legal Decisions," *Nature* 444 (2006): 664–665, 665.

4. Writer William Safire is credited with the term, defining it slightly over a decade ago as "the examination of what is right and wrong, good and bad about the treatment of, perfection of, or unwelcome invasion of and worrisome manipulation of the human brain." William Safire, "Our New Promethean Gift," remarks at the conference "Neuroethics: Mapping the Field," San Francisco, CA, May 13, 2002, the Dana Foundation, accessed September 4, 2012, http://www.dana.org/news/cerebrum/detail.aspx?id=2872.

5. Sam Harris, *The Moral Landscape: How Science Can Determine Human Values* (New York: Free Press, 2010), 2.

6. Tom Wolfe, "Sorry, but Your Soul Just Died," in *Hooking Up* (New York: Picador, 2000), 90.

7. David Dobbs, "Naomi Wolf's 'Vagina' and the Perils of Neuro Self-Help, or How Dupe-amine Drove Me into a Dark Dungeon," *Wired Science Blogs,* September 10, 2012, http://www.wired.com/wiredscience/2012/09/naom-wolfs-vagina-the-perils-of-neuroself-help/.

INDEX